THE GREEN WOODPECKER

Adult male Green Woodpecker. Kocsér, Hungary, June 2020 (RP).

THE GREEN WOODPECKER

A Natural and Cultural History of *Picus viridis*

GERARD GORMAN

PELAGIC PUBLISHING

First published in 2023 by
Pelagic Publishing
20–22 Wenlock Road
London N1 7GU, UK

www.pelagicpublishing.com

The Green Woodpecker: A Natural and Cultural History of Picus viridis

https://doi.org/10.53061/RYMI8301

British Library Cataloguing in Publication Data
A catalogue record for this book is available
from the British Library

ISBN 978-1-78427-436-8   Pbk
ISBN 978-1-78427-437-5   ePub
ISBN 978-1-78427-438-2   PDF

Cover photo: Green Woodpecker *Picus viridis* male in flight.
© Michel Poinsignon/naturepl.com

Typeset by BBR Design, Sheffield

Contents

Green Woodpeckers are invariably shy birds and studying them, especially around their nest, requires perseverance and patience. Ultimately, one must spend many hours with these birds to get to know them. Here, at a discreet distance from an active cavity, the author waits for a pair to arrive with food for their nestlings. Vértes Hills, Hungary, April 2022 (AK).

About the author

Gerard Gorman is an acknowledged authority on the Picidae (woodpeckers) having spent much of his life searching for, observing, listening to and studying these enthralling birds. He has written many papers and articles, and an unparalleled seven previous books on the family. *Woodpeckers of Europe: A Study of the European Picidae* (2004) is the only monograph devoted to all of the European species. *The Black Woodpecker: A Monograph on* Dryocopus martius (2011) is a comprehensive single-species account. The monumental *Woodpeckers of the World: The Complete Guide* (2014) is an acclaimed photographic review of all the species on the planet. *Woodpecker* (2017) explores both the natural and cultural history of woodpeckers worldwide. *Spotlight Woodpeckers* (2018) focuses on the four species that occur in Britain. Most recently *The Wryneck: Biology, Behaviour, Conservation and Symbolism of* Jynx torquilla (Pelagic Publishing 2022) has been described as 'the definitive work on this weird and wonderful bird'. Gerard currently lives in Budapest, Hungary, and is a founder member and current leader of BirdLife Hungary's Woodpecker Group.

Acknowledgements

Over the years, I have been privileged to meet, share experiences with and learn from some knowledgeable woodpecker enthusiasts. All of them, directly and indirectly, knowingly and unknowingly, have contributed to this book. I sincerely thank them all. The following, however, deserve special mention. Nigel Massen, David Hawkins and everyone at Pelagic Publishing were professional and supportive throughout the production of this book. My fellow woodpecker aficionado Daniel Alder was a superb sounding board, proposing many masterful edits. David Christie and Peter Powney greatly improved my text, making numerous valuable suggestions. Kyle Turner shared his in-depth knowledge of woodpecker sounds and created the spectrograms. Thomas Hochebner taught me much about woodpecker moult. Gergely Babocsay (Mátra Museum) and Tibor Fuisz (Budapest Museum) facilitated my visits to the bird collections of the Hungarian Natural History Museum. These generous people also helped in myriads of ways: Vasil Ananian, Korsh Ararat, Sanja Barišić, Leon Berthou, Taulant Bino, Mike Blair, Jean-Michel Bompar, Merijn van den Bosch, Ioana Catalina, Josef Chytil, Ricky Cleverley, Ármin Csipak, Péter Csonka, Tibor Csörgő, Tomasz Figarski, Kaspars Funts, Kai Gauger, Dimiter Georgiev, Keramat Hafezi, Paul Harris, Rolf Hennes, Erik Hirschfeld, Remco Hofland, Julian Hughes, Łukasz Kajtoch, Antal Klébert, Gábor Horváth-Mühlhauser, Thanos Kastritis, Chris Kehoe, Rolland Kern, Denis Kitel, Mati Kose, Serguei Kossenko, Tatiana Kuzmenko, Stephen Menzie, Karlis Millers, István Moldován, Killian Mullarney, Samuel Pačenovský, Nikolai Petkov, Tatiana Petrova, Mátyás Prommer, Dave Pullan, Borut Rubinič, Eldar Rustamov, Milan Ružić, Ken Smith, Domen Stanič, Daniel Szimuly, Ehsan Talebi, Dirk Tolkmitt, Josip Turkalj, Andreas Wenger, Volker Zahner and Bojan Zeković. Thank you, too, to Sergi Herrando of the European Breeding Bird Atlas (EBBA2) for allowing the use of the maps on pages 63 and 65. For kindly providing the photographs that enrich this book I am grateful to Vasil Ananian (VA), Vaughan and Svetlana Ashby/Birdfinders (VA/SA), Aurélien Audevard (AA), Fabio Ballanti (FB), Szymon Beuch (SB), Jean-Michel Bompar (JMB), Neil Bowman (NB),

Carlo Caimi (CC), Rob Daw (RD), Dimiter Georgiev (DG), Tomáš Grim (TG), Lisa Haizinger (LH), Thomas Hochebner (TH), David Hosking (DH), Terézia Jauschová (TJ), Gnanaskandan Keshavabharathi (GK), Antal Klébert (AK), Szabolcs Kókay (SK), Georges Olioso (GO), Dave Pearce (DP), Rudi Petitjean (RP), Bálint Stinner (BS), Maciej Szymański (MS), Elena Ternelli (ET), Kyle Turner (KT), Nick Upton (NU), Stephan Weigl (SW) and Phil Winter (PW). My own images are labelled (GG).

Gerard Gorman, Budapest, January 2023

Preface

… while ever and anon the measured tapping of Nature's carpenter, the
great green woodpecker, sounded from each wayside grove.

Sir Arthur Conan Doyle, *The White Company* (1891)

With its mostly green body plumage, vivid crimson crown, black 'Lone Ranger'
or 'Zorro' facial-mask and bright yellow rump – strikingly revealed when it flies
up and away from the ground – the Green Woodpecker *Picus viridis* is a stunning
member of the picid family. This is a bird that has adapted to live in both rural
and urban environments. It is often encountered while feeding on garden lawns,
in parks, pastures, even on sports fields and golf courses, and its unmistakable
'laughing' call means that it is well known to country and town folk alike. Indeed,
its 'laugh' is so recognisable that in rural England this was given a special name,
'yaffling', and the woodpecker itself became known as the 'Yaffle'.

Yet all of this does not mean that these woodpeckers are generalists that can
live anywhere. On the contrary, Green Woodpeckers are highly specialised birds.
In conservation terms they are special, too. They are a 'keystone species' as they
help shape the habitats in which they live and perform an important role for other
wildlife by providing tree cavities. Furthermore, they are an 'umbrella species'
as their conservation invariably confers protection on many other animals. They
are also an 'indicator species' in that their population status is indicative of the
biodiversity and health of the woodland and grassland habitats they frequent.

There is no substitute for watching woodpeckers in the wild, but if, for
whatever reason, you cannot regularly do so, then hopefully this book will
be of help. Whether you are already familiar with this species – if you are
an ornithologist, ecologist, citizen naturalist, birder, walker, forester, farmer,
gardener or a combination of any of these – or not, my aim in writing this
monograph remains the same: to take you a little deeper into the wonderful
world of the Green Woodpecker.

Finally, remember this: a world without woodpeckers would be a woeful
one. Do not take them, nor indeed any wildlife, for granted.

Adult male Green Woodpecker. Novo Yankovo, Bulgaria, December 2020 (DG).

FIGURE 1.1 An adult female Green Woodpecker in a village garden. Nógrád, Hungary, April 2022 (GG).

Chapter 1

Origins and Taxonomy

Woodpeckers are members of the Picidae, a cosmopolitan avian family in the order Piciformes. Globally they are the most widespread and the largest single family of the Piciformes, occurring from sea-level to high elevations on every continent except Oceania (Australia, New Zealand, Papua and islands east of there) and Antarctica. They are (not surprisingly) absent also from the Arctic and, more surprisingly perhaps, from Madagascar. Research on both genetics (molecular-sequence analysis) and morphology (structure and physique) strongly suggest that the closest relatives of woodpeckers are the honeyguides (Indicatoridae) of Africa and Asia. Other members of the order are the barbets of Africa (Lybiidae), Asia (Megalaimidae) and South America (Capitonidae), and the South and Middle American toucans (Ramphastidae), puffbirds (Bucconidae) and jacamars (Galbulidae) (Winkler 2015).

Evolution

The Piciformes have an exceptionally long history. It is believed that they began to evolve around 60 million years ago, in the Paleocene epoch. It is thought that the Picidae first evolved in what is now Europe and Asia after diverging from their close relatives about 50 million years ago (Sibley and Ahlquist 1990). If this is correct – and it is not yet known for certain from where and from what they evolved – it would make these birds one of the most ancient known avian forms. Woodpeckers as we know them now are probably akin to those that lived in the Pliocene epoch, around 5 million years ago (Winkler 2015).

Fossils

Fossils from birds that resemble what we recognise today as woodpeckers are scarce. Those unearthed in Europe reveal that woodpeckers were already present at the beginning of the Paleogene era (*c.*66–23 million years ago), represented by species of the tropical Capitonidae and the now extinct Zygodactylidae families. Fossils have been found in the Upper and Middle Pliocene deposits of Hungary, including one named *Picus pliocaenicus* (Kessler 2014). Fossils considered to be of *Picus viridis*, the Green Woodpecker, have been found in Austria, Croatia, Czech Republic, England, France, Germany, Hungary, Italy and Spain, and dated to the Lower, Middle and Upper Pleistocene and the current Holocene epoch (Kessler 2016).

Taxonomy

The evident morphological uniformity of woodpeckers customarily led taxonomists (and ornithologists) to categorise them primarily on plumage features. The development of taxonomy and phylogenetics of woodpeckers

FIGURE 1.2
An adult female Green Woodpecker. Kocsér, Hungary, November 2011 (RP). Fossils of this species dated from the Pleistocene have been unearthed in several countries across Europe.

has demonstrated, however, that most plumage colour patterns are unreliable for classification as they are routinely subject to convergence. In recent times, analyses of molecular data have revealed that the prevailing classification of these birds is far more complicated and definitely requires prudent revision (Winkler et al. 2014). Whatever the case, this topic and debate is ultimately beyond the scope of this book as it focuses upon a single species.

Classification

Picus viridis, literally 'Woodpecker green', was first described and classified by Carl von Linné (Linnaeus) in 1758, in the tenth edition of his *Systema Naturae*. The type locality was Sweden. In some of the historical literature this taxon also appears as *Gecinus viridis* owing to it being separated into the new genus *Gecinus* by Friedrich Boie in 1831 along with several other 'green' woodpeckers, where it remained for a time before being moved back to the *Picus*. Today, this woodpecker perches in the taxonomy of the animal kingdom as follows:

- *Kingdom:* Animalia (animals)
- *Phylum:* Chordata (vertebrates)
- *Class:* Aves (birds)
- *Order:* Piciformes (woodpeckers, honeyguides and allies)
- *Family:* Picidae (woodpeckers, piculets, wrynecks)
- *Subfamily:* Picinae (true woodpeckers)
- *Genus:* Picus
- *Species:* Picus viridis (Eurasian) Green Woodpecker

For more on the *Picus* genus and the closest relatives of Green Woodpecker see Chapter 5, Relatives.

A note on nomenclature

In the past, *Picus viridis* appeared in some books and lists as the 'Great Green Woodpecker'. In more modern times, two other English names are frequently used – Eurasian Green Woodpecker and European Green Woodpecker. These are useful to differentiate the species from other 'green woodpeckers' worldwide, such as Cuban Green Woodpecker *Xiphidiopicus percussus* (Cuba) and Little Green Woodpecker *Campethera maculosa* (Africa), which are not closely related to the Green Woodpecker of Eurasia. In the interests of brevity, the simple and widely used English name for the species, 'Green Woodpecker', is mostly employed in this book.

The subspecies

Green Woodpecker is polytypic, the majority of taxonomists and authors nowadays recognising three subspecies as follows: the nominate *Picus viridis viridis* (described by Linnaeus in 1758) which occurs in Britain, southern Scandinavia and continental Europe into the southern part of European Russia; *Picus viridis karelini* (described by von Brandt in 1841) in Italy, the southern Balkans, the Caucasus and on to the Kalibar Mountains, the Caspian region and Golestan in northern Iran; and *Picus viridis innominatus* (Zarudny and Loudon 1905) which is found as a disjunct population in north-east Iraq (Iraqi Kurdistan) and south-west Iran (western Zagros and Kordestan to central Fars Province) (Kaboli et al. 2016; Khaleghizadeh et al. 2017). For descriptions of these three commonly recognised subspecies see Chapter 3, Description and Identification.

'Zagros' or 'Mesopotamian' Woodpecker

Studies of the geographical distribution of the Green Woodpecker's lineage (phylogeography) have shown that genetic differences between populations across its range are slight (Pârâu and Wink 2021). However, some authors consider *innominatus* to be a distinct, separate species. Its isolated distribution in north-east Iran and south-west Iraq, as well as some morphological and genetic variations, have been put forward as different enough from the other two subspecies, which occur mainly in Europe, to grant it species status (Perktas et al. 2011; Perktas et al. 2015). A recent study in Iran also found a clear ecological-niche separation for *innominatus*, identified by strong differences in climate which affected vegetation (Elahi et al. 2020). The English names 'Zagros' and 'Mesopotamian' Woodpecker have even been proposed in the event of a split occurring (Perktas et al. 2011). Nevertheless, the above-mentioned differences have not been deemed by the major taxonomic authorities as sufficient to declare *innominatus* a separate species. Clearly, it is geographically the most allopatric (occurring in separate non-overlapping geographical areas) of all the Green Woodpecker subspecies, but more genetic research will be needed before it can genuinely be added to any birding checklists as a full valid species. To complicate matters further, *karelini* and *viridis* almost certainly overlap in range. They coincide in northern Italy along its borders with France, Switzerland, Austria and Slovenia, and the distributions of these two subspecies are also vague in the Balkans, for example in Bulgaria. Obviously, birds do not respect human-made national borders and racial intergrades surely must occur.

FIGURE 1.3 An adult male Green Woodpecker nominate subspecies *viridis*. Toulon, France, April 2020 (JMB).

FIGURE 1.4 An adult female Green Woodpecker subspecies *karelini*. Emilia-Romagna, Italy, August 2020 (ET).

Formerly claimed subspecies

In the past, other subspecies have been proposed, usually based on wing and bill lengths, overall body size and sometimes plumages patterns, but these are not generally accepted today as such differences are considered slight and attributed to clinal variations. They include '*pluvius*' in Britain, '*pinetorum*', '*virescens*' and '*frondium*' in Central Europe, '*pronus*' in Italy, '*romaniae*' in Romania, '*dofleini*' in the southern Balkans, '*saundersi*' in the Caucasus, and '*bampurensis*' in south-east Iran. This last was named after the Bampur River Basin in Baluchestan and was described on the basis of just two collected birds, which were said to resemble *innominatus* but were more sharply barred on the tail and wing and had barring extending over the whole of the lower breast, but there have been no observations for over a century and the whereabouts of the specimens are unknown (Khaleghizadeh et al. 2017).

Chapter 2

Anatomy and Morphology

Woodpeckers have an anatomy that complements their lifestyle. Form and function are correlated. Several comprehensive studies have demonstrated that they have evolved an array of adaptive bodily features that relate to their foraging and nesting behaviours (Bock 1999, 2015; Goodge 1972). There are, however, some subtle differences in how the various physical adaptations have evolved in different woodpecker families. The more arboreal species have the most robust bills, powerful excavation abilities and complex brain-protection features, while mainly terrestrial foragers such as the Green Woodpecker have the longest tongues and fewer shock-absorbing adaptations (Kirby 1980).

Woodpecker skulls are thick but relatively supple, as the bone is sponge-like in composition. The skull is positioned above the line of the bill, and the space between it and the brain is narrow and filled with relatively little cerebral fluid. All this means that the brain is not easily shaken, and potentially injurious vibrations arising from hammering on and into trees, for instance when excavating a nesting cavity, are diverted away from it. The pygostyle is a triangular plate of fused vertebrae at the end of the vertebral column. Also called the ploughshare-bone, because of its shape, it is flatter and larger in woodpeckers than in other birds and provides an anchor and support for the stiff shafts of the central tail feathers when they are pressed against a tree.

An adaptation that is perhaps unique to woodpeckers is an inwardly curved hinge of spongy tissue (the maxilla), which is located between the upper mandible and the front of the skull. It works to cushion and dissipate pressure on the brain when a bird hammers on or into hard surfaces. In Green Woodpeckers the tissue is less developed than in those woodpeckers that find prey mainly by pecking into trees (Bock 1999). It is also known that the anatomical arrangement of muscles and tendons in the cranio-cervical area, where the skull and spine connect, and in the ventral muscles of woodpeckers

FIGURE 2.1 Skeleton of a Green Woodpecker. Note the large decurved bill, long toes and claws and the ploughshare-shaped pygostyle bone (GG).

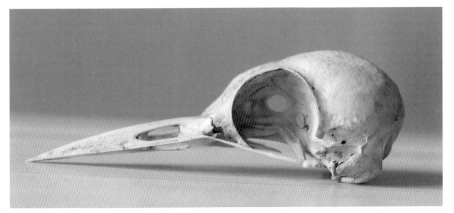

FIGURE 2.2 Green Woodpecker skull and 4 cm long bill. Item C.61.15.1., Hungarian Natural History Museum, Budapest (GG).

are vital in enabling them to drum on and peck into wood. Although all of these adaptations are advanced in the Green Woodpecker, they are even further developed in more arboreal species such as the Black Woodpecker *Dryocopus martius* (Böhmer et al. 2019).

All members of the woodpecker family have a particularly thick skin, which is an evolutionary outcome of their way of life. In the case of the Green Woodpecker, its tough skin provides some protection from wood splinters and from formic acid in attacks and bites by ants, which may get under its feathers. A thick nictitating membrane that closes over the eye protects it just before the impact of excavation work, although this feature is not at all unique to woodpeckers. The inner ear, too, resists injury by virtue of a thickened membrane.

The bill

Most woodpeckers have a straight, chisel-like bill, but as a species that often feeds on the ground and seldom drums the Green Woodpecker has evolved a comparatively long and decurved bill, which, though still robust, is also apt for digging into soil and turf and even through snow.

The nostrils are narrow slits and are covered by nasal bristles, specialised feathers that consist mostly of a rigid rachis. These function as a protective barrier: a filter that prevents the dust and debris produced when excavating wood from entering the nostrils and respiratory system.

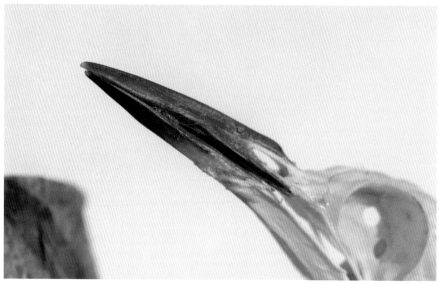

FIGURE 2.3 The impressive Green Woodpecker bill is long (this one around 4.5 cm) and decurved. This specimen from a private collection in Hungary (GG).

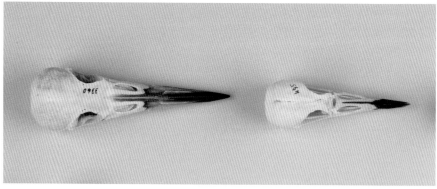

FIGURE 2.4 Comparison of two skulls and bills of Black (left) and Green (right) Woodpeckers. This particular Green Woodpecker bill is about 4 cm long. Within the Green's range, only Black has a larger bill. Items 3360 and 495, Natural History Museum, Vienna, Austria (GG).

Feet, toes and claws

Not all woodpecker feet are the same. Those of the highly arboreal species are subtly different from those of the 'ground woodpeckers' such as the Green Woodpecker. Yet, as in all woodpeckers, Green Woodpeckers have strong, deeply curved claws and short scaly legs with powerful hind muscles, all of which aid climbing and gripping (Bock 1959, 2015). Most woodpeckers (but not all, as some such as the appropriately named Three-toed Woodpecker

FIGURE 2.5 Coarse nasal bristles that cover the nostrils at the base of the upper mandible help stop wood dust and soil from getting up the woodpecker's nose (GG).

FIGURE 2.6 This male has fine wood debris on its head: various adaptations help stop such fragments from entering the eyes and nostrils. Kocsér, Hungary, August 2018 (RP).

FIGURE 2.7 A juvenile Green Woodpecker sits across a thin branch. Although woodpeckers in general are not regarded as 'perching birds', the zygodactylous arrangement of their feet is actually more suited to perching than to clinging to tree trunks. Kocsér, Hungary, June 2020 (RP).

Picoides tridactylus lack the hallux, the small first toe) have four toes on each foot and a 'zygodactylous' foot arrangement. This 'paired-toed' layout has two toes pointing forwards (digits 2 and 3) and two backwards (digits 1 and 4), with two toes on each side (Bock 1959). It is often referred to as the 'yoke arrangement' (zygomorphic means 'yoke-shaped'); however, this term is misleading as woodpecker feet are not strictly zygomorphic (Wenger 2021a).

Interestingly, the basic zygodactylous layout of the foot is not regarded as ideal for climbing, but rather for perching. It has evolved to optimise gripping. As a consequence, the hind toe (digit 1, the hallux) is significantly reduced in size in most woodpeckers (and, as already mentioned, even missing in some) and tilted to the side. To compensate for this lack of functionality (or total loss), digit 4 is moved to the rear to provide stability when climbing upwards. Digits 3 and 4 are the thickest and roughly the same length, and so provide strength when grasping branches. Furthermore, the zygodactylous alignment of the toes is not unique to woodpeckers as other members of the Piciformes also have it, as do, for instance, cuckoos (Cuculiformes), owls (Strigiformes) and parrots (Psittaciformes). Actually, a more precise term for how woodpeckers position their toes when climbing is 'ectrodactyl' (Bock 1959, 2015). As Green Woodpeckers tend to spend more time foraging on the ground than on tree trunks – thus horizontally rather than vertically – they probably employ the zygodactylous toe format more often than most of their sympatric relatives (those which occur within the same or overlapping geographical areas).

Notwithstanding the above, there are two anatomical features of the Green Woodpecker (and the *Picus* genus in general) that combine to form a truly remarkable foraging mechanism and deserve special mention: the tongue and the hyoid bones.

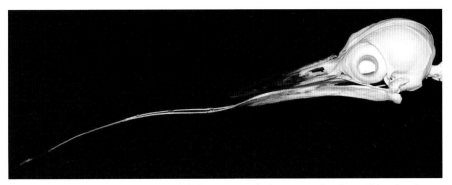

FIGURE 2.8 Skull, bill and remarkably long extended tongue of a Green Woodpecker (SW/LH).

The tongue

Woodpecker tongues have consistently intrigued anatomists. This astonishing structure was commented upon by such luminaries as Aristotle, Charles Darwin and Leonardo da Vinci, the latter famously noting that he should not forget to describe the tongue of the woodpecker. In *A History of British Birds, Indigenous and Migratory* (1837–52), William MacGillivray specifically described the Green Woodpecker's tongue, discussing its function, with detailed drawings, over five pages. This species' tongue is indeed particularly impressive. It is protrusible and not only proportionally longer but also flatter and broader at the tip than those of the more arboreal woodpeckers with which it is sympatric. It is typically around 13 cm in length and can be extended outside the bill by about 10 cm (Cramp 1985). As a general rule, the more terrestrial the woodpecker species, the longer and stickier is its tongue and the fewer sharp barbs there are on it. This is certainly true for Green Woodpeckers, which mostly use this retractable and supple tool, its tip rich in sensors, for lapping up ants rather than impaling wood-boring larvae and other invertebrates. The species also has a well-developed salivary gland, on average larger than those of other birds of a similar size, positioned below and behind the brain case (Leiber 1907). This coats the tongue with a viscous mucus to which ants and other prey adhere, rather than being impaled (Sielmann 1961).

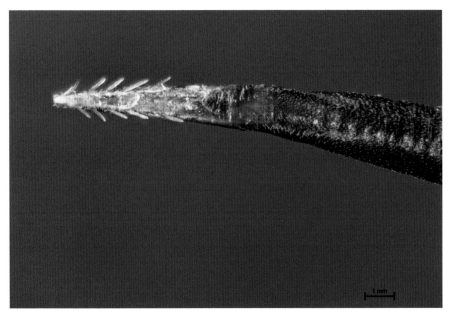

FIGURE 2.9 The magnified tip of a Green Woodpecker's tongue showing its barbs: relatively few for a woodpecker (SW/LH).

The hyoid apparatus

The hyoid is a bony but flexible structure that supports the tongue. It is attached to the muscles supporting the tongue, near the base of the upper mandible. In woodpeckers its two main bones (often referred to as 'horns') are longer than they are in most other birds, and when retracted they curve around and loop over the skull (Leiber 1907). When pushed forward, the hyoid apparatus forces the tongue out of the bill and helps control and manoeuvre it when it is inserted through narrow openings such as insect galleries. Long, flexible hyoid bones are a typical anatomical feature of terrestrial woodpeckers and reflect their foraging techniques and diets of mainly ground-dwelling ants (Short 1971). A further function of the hyoid bones is that, when they are withdrawn and stowed around the skull, they provide the brain with an additional layer of protective cushioning.

Although Green Woodpeckers, and the other members of the *Picus* genus, spend much time foraging on terra firma, and have an accordingly appropriate anatomy for this, it is worth noting that they are not alone among woodpeckers in this preference. There are many other picids around the world that regularly drop down from the trees to feed. Two, the Andean Flicker *Colaptes rupicola* of the Andes Mountains in South America and the aptly named Ground Woodpecker *Geocolaptes olivaceus*, which is only found in South Africa, Lesotho and Eswatini, have done so almost totally. They are highly terrestrial, much more so than Green Woodpecker. These two species

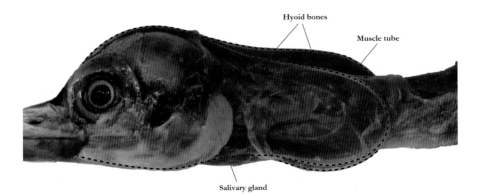

FIGURE 2.10 Skinned head of Green Woodpecker showing the hyoid apparatus, muscle tube and large salivary gland (SW/LH).

FIGURE 2.11 An adult male Green Woodpecker with soil in his bill after having used it to dig into the ground for prey. Kocsér, Hungary, November 2011 (RP).

have adapted to living in mostly treeless habitats such as rocky hillsides and montane grasslands, not only habitually searching for food on the ground, but even excavating their cavities in earth banks, gullies, road and rail cuttings, mud-brick ruins and abandoned buildings rather than in trees (Gorman 2014). A detailed examination of their anatomies would no doubt reveal even more incredible insights; however, that is for another time and perhaps other authors.

Chapter 3

Description and Identification

The Green Woodpecker is the second-largest woodpecker across its range. The Black Woodpecker is the biggest, at around 50% larger; its close relative the Iberian Green Woodpecker *Picus sharpei* is slightly smaller and the Grey-headed Woodpecker *P. canus* about 20% smaller. The vernacular name is not only descriptive but accurate, as the plumage of Green Woodpeckers is indeed mostly green, although there are areas of black and red and the rump is yellow.

Jizz

Green Woodpeckers appear stocky, sturdy and heavy, particularly when seen in their characteristic bounding flight (see Chapter 11, Movements and Flight). Flying female and juvenile Golden Orioles *Oriolus oriolus* might possibly be mistaken for Green Woodpeckers on brief views when they fly, as they too are greenish, but slimmer and smaller. Yet in most situations Green Woodpeckers can only realistically be confused with Iberian Green or Grey-headed Woodpeckers (see Chapter 5, Relatives). When on the ground Green Woodpeckers move with a hopping gait, often keeping a rather stiff, upright posture which can appear somewhat awkward. The tail is kept down, touching the ground and is never cocked. When scaling tree trunks, they move upwards with the feet together in abrupt, jerking movements, keeping their tail pressed against the trunk between each hop. Green Woodpeckers never move head-first down trees. Males and females both have a black 'burglar' facial-mask with a pale eye at its centre which creates, anthropomorphically speaking, a serious or stern look.

FIGURE 3.1 A Green Woodpecker in a typical upright pose on the ground. Note the 'severe' expression. Red in the black malar stripe (moustache) indicates a male. Toulon, France, May 2019 (JMB).

Measurements

The following figures are averages for adult birds based on the main ornithological handbooks and databases, and also measurements taken by the author of skins (specimens) in the Bird Collection of the Natural History Museum, Vienna, Austria and the Budapest and Mátra Museums of the Hungarian Natural History Museum.

- Length from bill-tip to tail-tip: 32 cm (range 30–36 cm).
- Wing length: 16.3 cm (males 15.8–17.2 cm, females 15.9–16.9 cm).
- Wingspan: 41 cm (range 40–46 cm).
- Tail: 10 cm (range 9.4–10.6 cm).
- Weight: 190 g (range 138–220 g).
- Bill length from the tip to the base of the skull, where it meets the upper mandible: 4.5 cm (range 4.0–5.3 cm).

Plumage colouration

Both sexes have a crimson crown from the forehead to the nape, which is often streaked grey, and are black on the face around the eyes and lores. The ear-coverts, neck sides and edge of the red nape are greyish, sometimes with a greenish tinge. The chin and throat are pale and occasionally have

FIGURE 3.2 A Green Woodpecker pauses while foraging amongst ground vegetation. Note the all-black malar stripe which indicates an adult female. Toulon, France, May 2016 (JMB).

greenish tones. The breast, belly and flanks are pale grey, often with a buffy hue. The flanks can be subtly barred or flecked with grey or pale green. The ventral area is dusky-grey, the lower belly speckled with grey-green-brown barring or chevrons. The mantle, back, scapulars and tertials are light green; the rump and uppertail-coverts are bright yellow when in fresh plumage, duller when worn. The outer tail feathers are black with green fringes dotted with buff-white spots. The undertail is usually more spotted than the upper, with the outer feathers greener with greyish barring. The outermost undertail feather is barred green and black. The primaries are brownish-black with cream spots, the secondaries olive faintly speckled white. The underwing coverts and axillaries are greenish-yellow and usually faintly barred with grey.

Bare parts

The Green Woodpecker's dagger-shaped bill is pointed at the tip, broad across the nostrils and gently curved on the culmen. It comprises over half the total skull length. The upper mandible of adults is dark grey; the lower mandible is yellowish from the base to about halfway. Juveniles may lack yellow on the bill. The upper mandible is also slightly longer than the lower one. The legs, feet and claws are grey. The irides are white, sometimes with a pinkish or bluish tinge, and the eyes are surrounded by a thin greyish orbital-ring of skin.

Sexual dimorphism

Green Woodpeckers are sexually dimorphic in plumage, although males and females differ only slightly. There are slight variations between the sexes in overall size and in bill and other measurements, but these are not obvious, significant or useful for identification in the field.

Adult male: The main plumage difference between the sexes is that males have red in a black malar stripe (sometimes referred to as the 'moustache') while that of females is all-black, lacking red.

FIGURE 3.3 Close-up of an adult male. Note the red in the malar stripe (which is only present in males) and the yellow on the lower mandible (which in present in both sexes). Kocsér, Hungary, July 2020 (RP).

Adult female: Besides the all-black malar, the black facial-mask of females may be slightly smaller, extending less beyond the eye. Females may also have more extensive barring on the flanks and sides of the breast, but this is too variable and not diagnostic.

Juveniles

Young birds are easily separated from adults until their first moult in late summer or autumn. They have bold cream-beige spotting and barring on the mantle, back, scapulars and wing-coverts. The coverts are sometimes so heavily spotted that they seem almost totally whitish. Juveniles are also mottled with cream, white or grey speckles and bars underneath, and the neck sides and ear-coverts are barred brown. They are also duller olive overall than adults and have blackish streaking on the body. This well-marked and scaly look is in part due to some feathers having dark sub-marginal lines.

The rump of juveniles is not as bright yellow as that of adults and can be barred white. Their red crown often has some orange or yellow feather tips at

FIGURE 3.4 Juveniles lack the black facial-mask of adults. Red and orange feathers in the developing malar stripe indicate that this bird is a male. Kocsér, Hungary, August 2018 (RP).

FIGURE 3.5 Juveniles often have all-dark irides and initially lack yellow on the bill. The red in the malar stripe indicates a male. Toulon, France, May 2020 (JMB).

the rear and is more mottled with grey, sometimes white, than that of adults. The wings and tail are barred grey-black, but the tail feathers and primaries are more or less like those of adults. Juveniles typically have dark, greyish irides, but eye colour is too variable to be used for ageing (Demongin 2016). For the sexing of juveniles see Chapter 4, Moult, Aging and Sexing.

Identification of subspecies

There are only slight differences in plumage and size between subspecies and in effect they are of little use in the field. For example, Green Woodpeckers in the east of their range are said to be generally larger than those in the west. The *karelini* subspecies found in Italy, the southern Balkans, Turkey and northern Iran tends to have duller green upperparts than the nominate subspecies, is overall duskier with its yellow and green areas of plumage less bright, and the feather tips of its cheeks and underparts are greenish (Demongin 2016). The calls of *karelini* do not differ appreciably from *viridis* (Fauré 2013). In south-west Iran and north-east Iraq, *innominatus* is the least known and studied subspecies. It is said to be much like *karelini* but paler green above, whiter on the face and underparts, and with a more markedly barred tail.

Chapter 4

Moult, Ageing and Sexing

Most birds moult all their plumage at least once a year, typically after the breeding season has finished. This is termed a 'post-breeding moult'. Most woodpeckers, on the other hand, do not renew all their feathers each year. Although they replace their primaries, they retain some primary coverts and secondaries during their post-breeding moult, so the renewal of the primary coverts is not linked to the corresponding primary, as it is in many species (Ginn and Melville 1983). Ultimately, the moult regime of the Green Woodpecker is, as with other woodpeckers, correlated to its breeding cycle and foraging habits. The order in which adults replace their feathers is coordinated and staggered so that they can continue activities without being hampered and are always able to fly. Adults complete one moult per year; there is no pre-breeding moult. Birds begin to drop feathers soon after the reproduction period has finished and this lasts into the autumn, falling between May and November (Cramp 1985; Glutz von Blotzheim and Bauer 1994).

Wing feathers

There are 10 primaries and 11 or 12 secondaries per wing. Primaries six (P6) and five (P5) are usually the longest. The outermost primary (P10) is reduced in size in adults but longer and broader (by up to 30%) in juveniles (Hansen and Synnatzschke 2015). The primaries are shed sequentially (ascendant) from the innermost (P1), and the secondaries from two centres, ascendant from S1, ascendant and descendant from S8 (Ginn and Melville 1983).

FIGURE 4.1 Open wing of a Green Woodpecker in autumn. Note the very small outermost primary. Styria, Austria. Sept 2016 (TH).

Tail feathers

There are 12 main tail feathers (rectrices) which are moulted using a distinctive strategy. Although Green Woodpeckers spend much time on the ground when foraging, they have, like all true woodpeckers, strengthened tail feathers which provide support when they scale and cling to vertical surfaces such as tree trunks. The central two rectrices are especially robust, with strengthened central shafts (rachides) and tapered tips. These are so important as props, which support the bird when it is in an upright position against a tree trunk, that they are replaced last, after all the others have been renewed (Jenni and Winkler 2020). Counting the central two feathers as R1, moult starts ascendantly in pairs from R2 to R6 until finally R1 (Ginn and Melville 1983). The central pair are also the longest, the outermost being much shorter than the others.

FIGURE 4.2 Feather-plate of the tail and body feathers of an adult Green Woodpecker. Sex unknown. These feathers were from the remains of a bird killed by an unknown mammalian predator in Lilienfeld, Lower Austria, in March 1993 (TH).

Moult patterns

The juvenile and adult plumages of most Eurasian woodpecker species are rather similar; however, those of the Green Woodpecker (and Iberian Green Woodpecker) differ significantly from those of their sympatric relatives. Quite why this stark difference in appearance exists is unclear.

Juveniles

The post-juvenile moult of the Green Woodpecker (and indeed all woodpeckers) is intriguing. For example, while in adults wing moult follows a known scheme, that of juveniles does not involve all the wing feathers. It is an extended, gradual process that continues until the autumn. In their first moult, they always replace their primaries and rectrices; lesser, median and greater coverts; and alula and most body feathers (Winkler 2013). The primaries attained after this first moult are longer than the ones they had when they fledged (Hansen and Synnatzschke 2015). As is typical for woodpeckers, the secondaries, inner coverts, and most primary coverts and tertials are usually retained, though in some cases one or two tertials and outer primary coverts are replaced (Miettinen 2002; Baker 2016).

Furthermore, this moult often begins while the young are still in the nesting cavity, which can be from late May to July depending upon location. To clarify: this is *before* fledging and *before* they have ever used any of their wing feathers for flight. The relatively long time that nestlings spend in the nest chamber allows them to replace their significantly reduced primaries when full-sized ones would serve little function. In a nutshell, it is likely that this moult pattern is an 'economy version', done in order to save time and energy (Bergmann 2018).

Adults

Both parents commence a post-breeding moult soon after their chicks have fledged, usually between June and November (Ginn and Melville 1983). Beginning with the wings, most adults renew all of their plumage, but some birds may retain a few primary coverts and very occasionally some secondaries (Miettinen 2002; Winkler 2013). The body feathers are the last to be dropped. This moult is completed in the autumn, the time depending upon the region and when breeding finished; birds in the south generally finish earlier than those in the north.

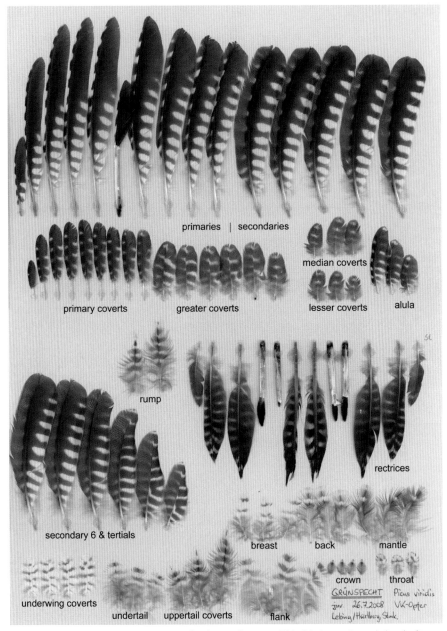

FIGURE 4.3 Feather-plate of a juvenile. Sex unknown. This bird was found dead after being hit by a vehicle in Hartberg, Austria, in July 2008 (TH).

FIGURE 4.4 Green Woodpecker in autumn. This bird's moult from juvenile into adult plumage is almost complete. Ócsa, Hungary, September 2021 (GG).

Sexing

Adults

As outlined in Chapter 3, the Green Woodpecker is sexually dimorphic. In the hand, as in the field, males and females can be separated by observing the colour of the malar stripe. Crown colour is not relevant. In this it is like its close relative the Iberian Green Woodpecker, but unlike the Grey-headed Woodpecker. Adult male Green Woodpeckers have red malar stripes bordered by black, while those of adult females are plain black.

Juveniles

Most juvenile males have some reddish, orange or gold in their (often faint) greyish-black malar stripes. Juvenile females lack these coloured feather tips and have a greyish malar stripe which, again, can be indistinct. Some males can also lack these colours, however, until they begin to appear during their first moult in the summer. Hence, malar colour in juveniles is not a dependable way of sexing individuals until the overall moult is advanced, and caution in sexing juveniles using this feature is advisable before this time (Demongin 2016).

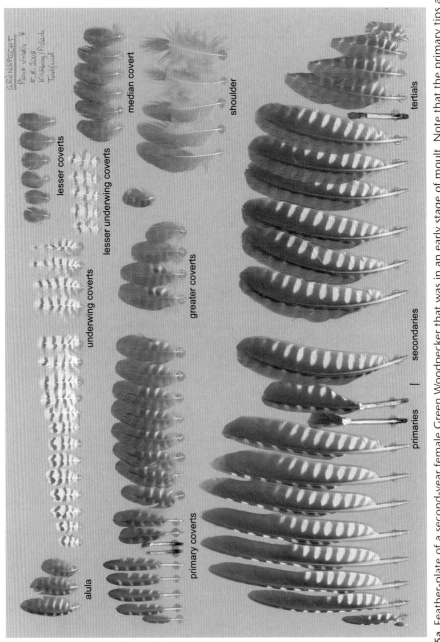

FIGURE 4.5a Feather-plate of a second-year female Green Woodpecker that was in an early stage of moult. Note that the primary tips are bleached and that the bird still had some juvenile body feathers, tertials, wing-coverts and alula. NB: this bird was found dead of unknown causes in Kirchberg, Pielach, Lower Austria, in August 2008 (TH).

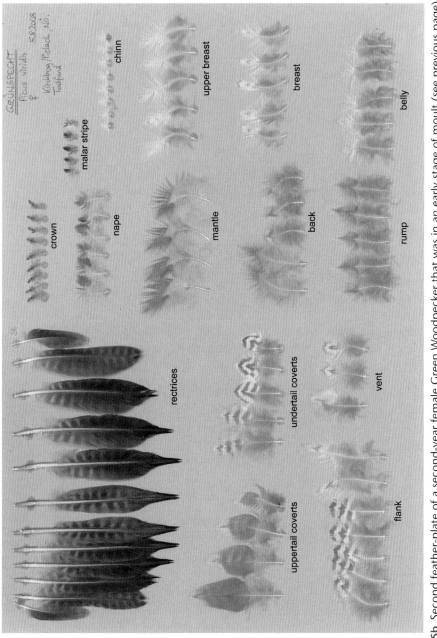

FIGURE 4.5b Second feather-plate of a second-year female Green Woodpecker that was in an early stage of moult (see previous page).

FIGURE 4.6 The sex of this juvenile is unclear, as both males and females may lack red in their developing malar stripes. Kocsér, Hungary, June 2020 (RP).

FIGURE 4.7 A juvenile male Green Woodpecker: note the red feathers in the faint malar stripe. The bluish tinge to the iris is not related to sex nor age. Styria, Austria, July 2014 (TH).

FIGURE 4.8 This individual is possibly a juvenile female as there is no red in the malar stripe, but as some juvenile males may also lack red there, sexing using this feature is ultimately unreliable. Styria, Austria, July 2011 (TH).

Ageing

As their plumages are so different, separating adults and juveniles in the field is straightforward. However, ageing Green Woodpeckers in the period from their first autumn to first spring, when they are moulting into full adult plumage, can be difficult. In that transitional period, markings on the coverts are one of the best ways of ageing Green Woodpeckers, but realistically these can only be seen in the hand (Mann 2016). All in all, probably only two age groups can be easily recognised: juveniles and full adults. Juveniles are unmistakable. Their underparts are white, cream or greenish-white, and heavily spotted, flecked or barred with brown; the upperparts are green with white spots. The tertials have pale tips and barring and are duller than in adults. The primary coverts are narrow. They lack the black facial-mask of adults and their irides are initially greyish-brown but soon become paler (Baker 2016). Following a partial moult, birds in their first autumn through to their first spring largely resemble adults. Their retained tertials generally have pale bars across the shafts and white tips. Any new tertials look fresh, greenish and contrast with the worn retained ones. As mentioned above, markings on the retained coverts are key to ageing. Retained primary coverts are pale, often greyish-green and

FIGURE 4.9 Open right wing of a juvenile Green Woodpecker. Note the mottled plumage and streaked face lacking a black mask. Red in the malar area indicates a male. Styria, Austria, July 2014 (TH).

with noticeable white barring. Any new primary coverts contrast sharply with worn ones. In addition, the facial-masks of these birds are mottled and patchy rather than totally black.

Adults have plain black facial-masks and unspotted body plumage. The tertials are plain greenish-buff, lacking pale tips and barring. The primary coverts are broad and rounded, olive with a conspicuous yellowish-green leading edge, and the distal bar running along the shaft has a white trailing edge. The presence of several generations of primary coverts indicates an adult bird (Winkler 2013). Adults have white irides, sometimes with a pinkish or bluish tint (Baker 2016).

Aberrants

Atypically plumaged Green Woodpeckers, such as birds with dark crowns lacking red or with pigment conditions such as albinism, leucism or melanism, are rarely observed (Chatfield 1970). Individual aberrant birds with greyish or brownish feathers, especially on the mantle, are more frequent (Demongin 2016). In some regions, birds that have unusual plumage may be the result of hybridisation with other species in the *Picus* genus (see Chapter 5, Relatives).

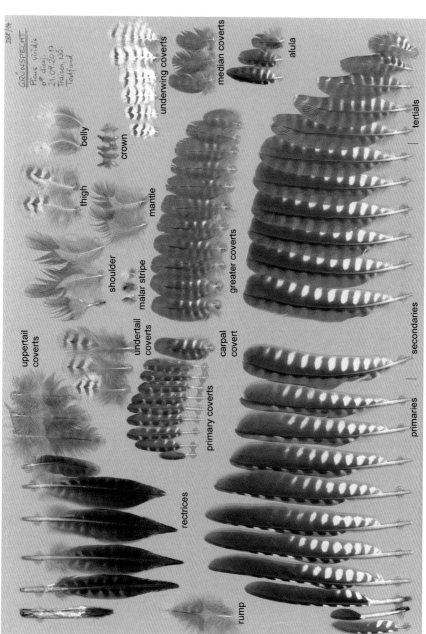

FIGURE 4.10 Feather-plate of a juvenile male Green Woodpecker that had nearly completed its body moult. Note that the tertials are barred white and all the primary coverts are the same generation. NB: this bird was not 'collected' but found dead of unknown causes in Traisen, Lower Austria, in September 2017 (TH).

Longevity

The oldest known Green Woodpeckers, documented from ringing (banding) recoveries, are a bird of 8 years old in Hungary (Török 2009), one of 12 years 10 months in the Czech Republic (Cepák et al. 2008), and in Britain a road-killed bird aged 15 years 1 month (Fransson et al. 2017).

Chapter 5

Relatives

All the species in the *Picus* genus are found in the Old World, across a wide area from Britain, Scandinavia, continental Europe and North Africa eastwards to India, China, Japan and South-East Asia. The base plumage colour of most is green. All are sexually dimorphic, males mainly differing from females in the extent of red they have on the crown or face (Gorman 2014). Several of these woodpeckers are chiefly terrestrial foragers that predominantly feed on ground-living ants in open areas, rather than in closed wooded habitats. Nevertheless, all excavate nesting and roosting cavities in trees. The precise number of species in this Eurasian genus is debated, with 13 (Gorman 2014; Winkler 2015; Clements et al. 2021), 14 (Gill et al. 2022) and 15 species (del Hoyo and Collar 2014) recognised.

The latter checklist includes the following: Crimson-winged Woodpecker *Picus puniceus* (SE Asia), Lesser Yellownape *Picus chlorolophus* (Indian subcontinent, SE Asia), Japanese Green Woodpecker *Picus awokera* (Japan), Red-collared Woodpecker *Picus rabieri* (S China, Laos, Vietnam), Streak-throated Woodpecker *Picus xanthopygaeus* (Indian subcontinent, SE Asia), Laced Woodpecker *Picus vittatus* (SE Asia), Streak-breasted Woodpecker *Picus viridanus* (SE Asia), Grey-headed Woodpecker *Picus canus* (Eurasia), Black-naped Woodpecker *Picus guerini* (Indian subcontinent, China, SE Asia), Sumatran Woodpecker *Picus dedemi* (Sumatra), Black-headed Woodpecker *Picus erythropygius* (SE Asia), Scaly-bellied Woodpecker *Picus squamatus* (Central Asia), Iberian Green Woodpecker *Picus sharpei* (Spain, Portugal, SE France), Levaillant's Woodpecker *Picus vaillantii* (Morocco, Algeria, NW Tunisia) and, of course, the main subject of this monograph.

From the above-listed species only two are sympatric with Green Woodpecker: Iberian Green and Grey-headed Woodpeckers, which overlap in range in some parts of Europe. Levaillant's Woodpecker occurs in north Africa

FIGURE 5.1 The Laced Woodpecker is a South-East Asian member of the *Picus* genus. This bird is an adult male. Malaysia, September 2016 (VA/SA).

FIGURE 5.2 Like its Eurasian relative, the Streak-throated Woodpecker often feeds on the ground. This female in Tamil Nadu, India, is taking termites (GK).

and does not coincide geographically with any other *Picus*. Two species, Iberian Green and Levaillant's Woodpeckers were previously regarded as conspecific with the Green Woodpecker. These close relatives have been treated in some detail in recent molecular studies (Fuchs et al. 2008; Pons et al. 2011, 2019; Perktas et al. 2011).

Iberian Green Woodpecker

This monotypic species occurs widely in Portugal and Spain, as well as in southern France on the northern slopes of the Pyrenees and north through the Hérault département to the Rhône Delta (Olioso and Pons 2011). So, despite the English vernacular name 'Iberian', the Pyrenees Mountains do not actually form a strict barrier that separates the Iberian Green Woodpecker in the south, in Spain, from the Eurasian Green Woodpecker in the north, in France. The Iberian Green Woodpecker has long been considered somewhat different from the Eurasian Green Woodpecker, and in recent times detailed studies on the genetics of the two have revealed that they are indeed separate species. They were officially split after systematic studies on plumage, morphology and genetics (Perktas et al. 2011; Pons et al. 2011; IOC 2017). Further research,

FIGURE 5.3 An adult Iberian Green Woodpecker on the ground in a classic *Picus* pose. As with Eurasian Green Woodpecker, the red malar stripe indicates a male. Madrid, Spain, November 2017 (RD).

which investigated the COI-5P region (an mtDNA marker widely used for species diagnosis), revealed significant divergence between the lineages of the two and thus left little doubt that they are genetically distinct (Arànega et al. 2020). It has been estimated that these lineages split about 0.7–1.2 million years ago, probably when they became isolated by geographical barriers that heightened allopatric divergence (Pons et al. 2011).

Structurally, the Iberian Green Woodpecker is on average slimmer and slightly smaller than the Eurasian Green, at 30–34 cm in length. Plumage-wise, it differs from its relative in lacking a black facial-mask, though some individuals can be rather dusky in that area, and in having unbarred undertail-coverts. Males have a mostly red malar with a thin, incomplete black edge. Females have a greyish malar. Its advertising call is structurally similar to that of the Eurasian Green and is often hard to differentiate as it consists of only one dominant frequency, while Eurasian Greens may 'yaffle' with a single dominant frequency along with the more typical double version (Fauré 2013, 2021). Generally, the Iberian Green's advertising call is faster, higher pitched and not as throaty, often somewhat variable and to some ears recalls the whistling of a Grey-headed Woodpecker (Gorman 2004). Local Spanish names for this woodpecker such as *Caballito*, meaning pony or little horse, refer to the whinnying nature of its main vocalisation.

Grey-headed Woodpecker

Sometimes called the Grey-faced Woodpecker, this species is very widespread over Eurasia, occurring from continental Europe and Scandinavia across the boreal zone to the Far East and northern Japan. The taxonomy of this polytypic species is under review. Some of the subspecies across its huge range may well be distinct species, especially those in South-East Asia (Gorman 2014). Some of these do indeed differ greatly in plumage colour and also in the habitats that they use, but so far, few genetic (molecular-sequence analysis) studies have been carried out on this group. The nominate subspecies *canus* is sympatric with Green Woodpeckers in many areas of Europe but is generally a more northerly and boreal species than its congener. The two species also occasionally hybridise. Although there is overlap in the habitats that they use, Grey-headed Woodpeckers tend to be found more often in closed forests than Green Woodpeckers (Winkler et al. 1995; Gorman 2004). Grey-headed is also less likely to be seen in urban environments. In addition, although obviously a *Picus* species that forages on the ground, Grey-headed is not as dependent upon terrestrial ants as its relative, feeding more often in trees and hence consuming a broader range of prey (Blume 1996). At 26–32 cm in length, it is visibly slimmer and smaller headed than the Green Woodpecker, lacks a black facial-mask and red crown and has only a slim black malar stripe. Males

FIGURE 5.4 A Grey-headed Woodpecker about to be released after having been ringed. The lack of red on the forecrown indicates a female. Ferencmajor-Naszály, Hungary, September 2020 (GG).

have a red forecrown patch, which can sometimes be difficult to discern, while females show no red at all. Hence, in most situations the two species should not be mistaken for each other. Its advertising call is a characteristic plaintive, whistling series of 6–10 piping *poo* or *pew*, or harsher *koo* notes that fall in pitch, and is obviously different from the 'yaffling' of Green Woodpeckers. This species is a much more frequent and avid drummer, both sexes producing rolls of typically 1–2 seconds long with 20–40 strikes per second (Gorman 2014). As the Grey-headed Woodpecker occurs over a very large geographical area, across which it is believed to be increasing in number, its International Union for Conservation of Nature (IUCN) Red List status is categorised as Least Concern (BirdLife 2022b).

Levaillant's Woodpecker

Also called Levaillant's Green Woodpecker, Maghreb Green Woodpecker, North African Green Woodpecker and Algerian Woodpecker, this is another monotypic species. It was first described and named *Chloropicus vaillantii* in 1847 by Malherbe from a specimen in Algeria, but since then its taxonomic position has fluctuated between species and subspecies (Winkler et al. 1995). Despite being traditionally treated as a subspecies of *Picus viridis*, Levaillant's is geographically separated from its Eurasian relative, being endemic to the Maghreb region and hence the only *Picus* on the African continent. Therefore, unlike Iberian Green and Grey-headed Woodpeckers, it does not coincide with the Green Woodpecker anywhere and consequently hybridisation has never been documented. These factors and strong morphological, ecological and more recently studies which established genetic differences, finally led to it being given species status. It is now accepted as such by all the main taxonomic authorities. Furthermore, this woodpecker is considered to be only distantly related to other *Picus* species populations, having diverged from them some 1.6–2.2 million years ago (Pons et al. 2011).

Levaillant's Woodpecker is smaller than the Green Woodpecker, at 30–32 cm in length. Visually it differs most obviously from its Eurasian and Iberian relatives in that males as well as females lack red in the malar stripe, this area being black with a bold white line at the top. Females are also unique among this group of relatives in having red only on the nape and sides of the rear crown. Juveniles are less streaked and spotted than juveniles of its nearest European relatives. The advertising call of Levaillant's Woodpecker is a more musical, higher pitched rapid series of whistling *pee-pee-pee* notes, not as 'yaffling' or mocking as that of Green Woodpecker, and it drums in the true

FIGURE 5.5 An adult male Levaillant's Woodpecker. Females have red only on the rear crown. Unlike Eurasian Green, both sexes of this species have all-black malar stripes. Morocco, April 2015 (AA).

sense and more frequently (Gorman 2014). Although its population size and trends are unknown, and it is somewhat localised across its range, Levaillant's Woodpecker is not uncommon locally and does not seem to be faced with any significant threats. Therefore, its IUCN Red List status is categorised as Least Concern (BirdLife 2022b).

Picus hybrids

Occasionally Green Woodpeckers interbreed with Iberian Green Woodpeckers, for example in Languedoc-Roussillon in southern France (Pons et al. 2019), and with Grey-headed Woodpeckers across Europe, including records from Belgium (Schmitz and Dumoulin 2004), Germany (Ruge 1966; Berger 1990;

Südbeck 1991; Bird and Südbeck 2004; Südbeck and Brandt 2004), Poland (Dmoch 2003; Czechowski and Bocheński 2012; Ławicki et al. 2015), Russia (Ivanchev 1993; Friedmann 1993, 2011) and Sweden (Salomonsen 1947). Some mixed pairings may involve a pure-species parent and a hybrid parent resultant from a previous mixed-pair breeding. Deciding upon and defining the rules for how to visually identify a cross is fraught with difficulties. The morphological features present in such individuals vary from those clearly inherited from one or both of their parents to those that are not present in either parent nor in intermediary forms. Generally, those hybrid offspring which are relatively easy to identify are those that show an obvious mixture of features inherited from both parents. Nevertheless, care should always be taken with unusual birds before designating them as hybrids, as individual variation and atypically plumaged individuals may occur. Likewise, although an unusual call or drum that sounds in some way different may draw an experienced observer's attention to a bird, the sounds made by these two species are probably too variable to use as a basis for diagnostic hybrid vocalisation.

Green × Iberian hybrids

There is contact and hybridisation between Eurasian Green and Iberian Green Woodpeckers in southern France, in a zone of about 245 km wide along the Mediterranean coast. Morphologically intermediate birds, which typically show various blackish and greyish hues on the face, are seen there (Olioso and Pons 2011). A study in the area between Béziers and Montpellier found several individuals that carried DNA from both species, and this put the species status of the Iberian Green Woodpecker in doubt. However, this hybridisation was shown to occur only in a secondary contact zone, and the study concluded that genes do not readily flow between the two (Pons et al. 2019).

Green × Grey-headed hybrids

Mixed plumage features of Green and Grey-headed Woodpeckers usually involve the colour and extent of the malar area and nape and/or the amount of black on the face. Indeed, a detailed review of hybrids of these two species in Poland found that the most reliable plumage feature was a black malar stripe that was narrower than it is on *viridis* but broader than on *canus*, and which was separated from the black facial-mask and lores (Ławicki et al. 2015).

In addition to the above details, the review found that dusky rather than black feathers below the eye also pointed to a hybrid. This facial pattern

FIGURE 5.6 This male was confirmed as a *Picus viridis* × *sharpei* cross after genetic analysis. Plumage-wise, note the dark facial area which is intermediary: too restricted and greyish for *viridis* but too extensive and blackish for *sharpei*. Hérault, France, June 2009 (GO).

FIGURE 5.7
This female was also confirmed as a *Picus viridis* × *sharpei* cross. As with the male above, the dark facial area is intermediary in colour and extent. Hérault, France, June 2010 (GO).

FIGURE 5.8 A *Picus viridis × canus* cross. Note that the crown is all-red, as is typical for Green Woodpecker, but only the lores are black with most of the orbital area grey. Bytom Miechowice, Upper Silesia, Poland. March–May 2012 (SB).

FIGURE 5.9
Another view of the same *Picus viridis × canus* cross shown above in figure 5.8. Bytom Miechowice, Upper Silesia, Poland. March–May 2012 (SB).

recalls that of an adult female Iberian Green Woodpecker (although there is no suggestion that this species was involved in these hybridisations in Poland). A feature often, but not always, found on hybrids, which is not present in either parent species, is a greyish or blackish patch near the rear of the red nape. Another example involved a hybrid male *Picus* individual with the shape and size of a Grey-headed Woodpecker but several plumage characteristics of a Green Woodpecker, which was seen and photographed in 2003 in Belgium, associating with a normal female Green Woodpecker. This individual had red on the crown and nape, which was more than is typical for Grey-headed, but less defined and more reduced than on a typical Green. Its malar was black and large, and it had a black facial-mask bordered by grey, a greyish-green chin with faint black specks and grey ear-coverts, cheeks and neck. Most of the underparts were greenish-buff, with greyish barred upper flanks and unbarred lower flanks. The outer feathers of the tail were slightly barred on the inner-web but unbarred on the outer-web. The undertail coverts were irregularly barred brown but with unbarred tips. The bill was a dull yellow and shorter and more pointed than for a Green Woodpecker. This Belgian bird was heard calling and was described as 'too sweet and soft' for Green but not as musical as Grey-headed. In total, four cases of interbreeding were encountered in eastern Belgium between 1986 and 2003, and in every case the female was a Grey-headed Woodpecker (Schmitz and Dumoulin 2004).

Hybrids or aberrants?

A Russian study on this subject found that although mixed pairs of Green and Grey-headed Woodpeckers were sometimes seen, they often did not breed successfully (Friedmann 2011). The authors thought that incompatible courtship behaviour was the reason that most mixed pairs failed to nest. Another theory was that infertility might also have been involved (Ivanchev 1993). Hybrids were, however, sometimes encountered and therefore successful breeding was presumed to have taken place. In addition, Green Woodpeckers with atypical plumage occasionally occur that are not hybrids, but simply aberrant individuals. Clearly, when unusual individual Green Woodpeckers are observed in areas where Grey-headed Woodpeckers and Iberian Green Woodpeckers are totally absent, such as in the British Isles, interbreeding cannot be the cause of birds having atypical plumage.

Chapter 6

Communication

Green Woodpeckers communicate with each other using four basic methods: simple calls, advertising calls, drumming and tapping. Both males and females use all four methods, with no obvious sex-related differences evident. The differences between various calls lie in their structures and functions, although there is some overlap. Almost all calls consist of repeated notes, from the simple two- or three-note alarms, given when the birds are startled or agitated, and the calls used to maintain contact with breeding partners and with young, to the longer and more complex advertising call. This 'laughing' call is used during courtship displays and to proclaim or defend a home range (see Chapter 12, Breeding) and is normally clearly defined. As with most woodpeckers worldwide, a general repetition of almost identical notes is a feature of vocalisations. Although they cannot really be described as good singers (at least from an anthropomorphic perspective), the advertising call of the species is certainly not unpleasant to human ears, being – perhaps – one of the most musical of any in the picid family. In the 1950s and 1960s, some detailed studies of the sounds made by the species were carried out, with six calls and two instrumental signals described (Blume 1955, 1961). More recently, individual variation within all call types have been described, but these do not appear to be regionally significant (Turner et al. 2022).

Drumming and tapping are non-vocal means of communication. Woodpeckers produce these ritualised instrumental, mechanical sounds by knocking on trees, and sometimes other surfaces, with their bill. In the avian world, drumming is unique to woodpeckers, although not all species drum. It is important to clarify that drumming and tapping are distinct from the incidental noises made during cavity excavation and foraging, which are not methods of communication but rather simple by-products of those activities. When drumming and tapping, woodpeckers do not create holes in the surface

FIGURE 6.1 A calling adult male Green Woodpecker. There are no obvious differences between calls made by the sexes. Bristol, England, March 2018 (RD).

they are beating upon. Instrumental signals are less complex and typically low frequency compared with vocalisation. Green Woodpeckers are not often heard drumming and tapping, since the sounds they produce are soft, quiet and hence easily overlooked by humans. Nevertheless, they probably constitute an important intimate communication between breeding pairs (Turner et al. 2022).

Spectrograms

Some of the descriptions that follow are accompanied by spectrograms. These graphs are a convenient visual means of depicting and describing recorded sounds. By examining spectrograms, variations in the calls made by different woodpecker species – and potentially individuals of the same species – can be compared. Measurements of call parameters from these graphs, for instance of cadence, can be made and this helps to identify patterns in the structures of the sounds. These patterns may also help with identification when remote recording surveys in dense forest are carried out. Spectrogram images of vocalisations provide an alternative to using words (mnemonics and phonetics which are rather subjective) as a means of describing them. In these graphs the frequency of a sound is plotted against time, with time shown on the horizontal x-axis in seconds and pitch (frequency) on the vertical y-axis in kilohertz (kHz). Amplitude is shown by tonal intensity.

Calls

The Green Woodpecker's repertoire of calls consists mostly of simple, repeated notes with the stress typically on the first note and a gradual deceleration thereafter. The most commonly used call, used in various situations, is a repeated series of *tiew*, *kew* or *teuk* notes which slow and either drop slightly in pitch or remain level.

Advertising call (song)

The loud, clear, forceful, advertising call of the Green Woodpecker serves a similar function to those of songbirds. Therefore, it might reasonably be described as the woodpecker's 'song'. That term, however, somehow does not seem suitable for a woodpecker, and hence few researchers use it (but see Fauré 2018, 2021). In whatever way it is described, this vocalisation is clearly more ritualised than the other calls, having a more complex and variable structure. Indeed, it is unmistakable and has been described as the most familiar and easily identifiable vocalisation of the species (Glutz von Blotzheim and Bauer 1994). It mainly functions as a territorial declaration but is also important in mate attraction, pair-bonding and perhaps to keep contact with recently fledged young. Green Woodpeckers do not only vocalise during the spring courtship period, but also in summer and autumn, though usually less

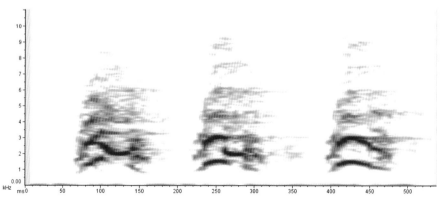

FIGURE 6.2 Regular call. Three *kew* note shapes in a single call. Recording made in Dorset, England, June 2005 (KT).

vociferously. It is likely that calls outside the breeding season serve as confirmations of home-range occupancy and to the fitness of the caller and thus, ultimately, to enhance breeding success in the following spring (Reichholf 2006). Unpaired birds vocalise more often.

The advertising call is not learned, is produced by both sexes and can be heard all year round, with latitude, local conditions and weather all influencing when individuals sing. Not surprisingly, however, the most intense period of calling is in spring, mainly from February to April (Blume 1962) but often into May. It is characterised by variables of length, pitch, emphasis, speed of delivery and degree of repetition. It is composed of a series of single notes, generally around 13, with a range of 2–34 and an average speed of 5.9 per second. The tempo varies, but most commonly decelerates, usually with a

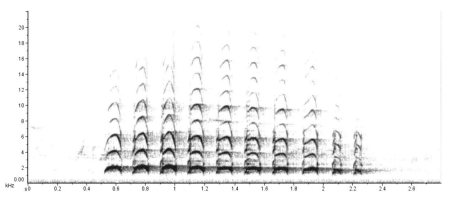

FIGURE 6.3 Advertising call. This 'yaffle' shows simple inverted 'U' notes plus two smaller, softer and sharper notes at the end, 10 notes at 5 per sec. Recording made in Dorset, England, February 2007 (KT).

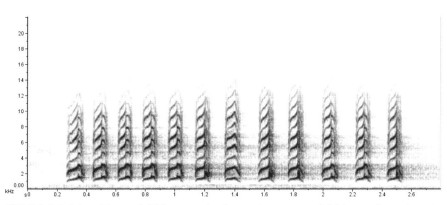

FIGURE 6.4 Advertising call. This example shows sharp notes of fairly consistent length and shape but with a clear deceleration, 12 notes at 4.5 per sec. Recording made in Lot, France, March 2005 (KT).

distinct ending but occasionally fading away. The rate of delivery of the advertising call also varies between individual birds. In words it might be rendered as *klu-klu-klu-klu-klu* or *kleu-kleu-kleu-kleu-kleu* or *kew-kew-kew-kew-kew*, or perhaps *plue-plue-plue-plue-plue*. Some individuals include shrill notes in this call, not unlike the alarms of some shorebirds. In terms of pitch there is a fairly consistent drop throughout the advertising call of most individuals of 0.3 kHz from the fundamental frequency of the initial note (Turner et al. 2022). An acoustic analysis of this call from one bird in the Netherlands found that the most dominant frequency component was 1,854 Hz and that the individual components were largely identical (Verboom 2019).

Contact call

A mated pair often exchange simple, soft, single or double *kuk*, *cook* or *kik* contact notes.

Flight calls

Flight calls composed of loud *kjaek* notes are made all year round. They are generally faster than calls made when birds are perched, but slow down as the flight ends. They are used to indicate a change of position and are a form of contact call made in flight.

Alarm calls

Alarms include a shrieking, shrill, high-pitched, loud, rushed *kyu-kyu-kyuk* and *kjuk-kjuk-kjuk-kjuk*, stressed on alternate notes. When threatened by an aerial predator, Green Woodpeckers typically utter a loud, panicky, cackling alarm call.

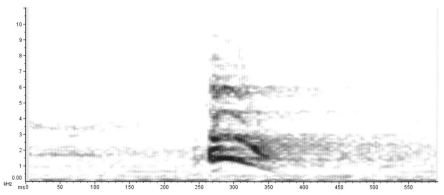

FIGURE 6.5 Contact *kuk* call. This was a bird emerging from its roost hole and responding to its mate's advertising call. Recording made in Wiltshire, England, January 2006 (KT).

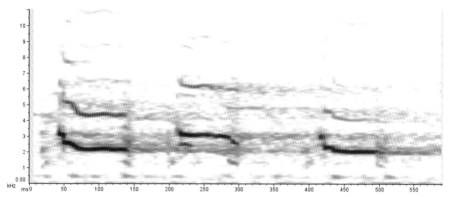

FIGURE 6.6 Flight call. This example shows a higher pitched second note; the middle of a series of three calls. Recording made in Lot, France, March 2008 (KT).

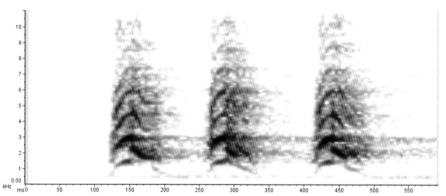

FIGURE 6.7 Alarm call. Forty-eight calls were given in seven minutes by this bird when it was disturbed near the nest. The call-rate slowed during the sequence and calls ranged from 3 to 6 notes. Recording made in Dorset, England, June 2001 (KT).

It might be transcribed as *kju-kju-kjuk* or *kjack-kjack-kjack* and is similar to the alarm call of a Grey-headed Woodpecker. Excitable, *Accipiter*-like, rather harsh *kek-kek-kek*, sharp *kik-kik-kik* and single or double *kewk* or *kyak* are also made.

Threat calls

A series of repeated, simple, evenly spaced, squeaky *kjaik* notes are given in various situations. For instance, by a pair when agitated near the nest owing to disturbance; in response to the advertising call of a mate, when arriving to feed young; and especially as a threat during encounters with rivals. The call may not be loud, but birds usually raise their red crown feathers when making it. A very similar call is also sometimes made in displays and has been likened to the rubbing of a window-pane with a damp chamois leather (Cohen 1946).

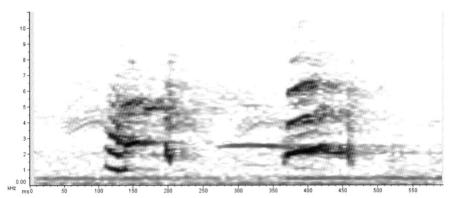

FIGURE 6.8 *Kjaik* threat call. First two notes from a call containing 34. The opening note is obviously sharper at the start. Typically given by adults when in close contact with juveniles. Recording made in Dorset, England, July 2003 (KT).

Courtship calls

Thin, soft *peeuw* and *pweep* calls are uttered in series, often over a dozen, during courtship, sometimes made by a mated pair in a duet.

Calls by the nest

Soft, intimate clucking *guk-guk* or *gluk-gluk* are uttered near the nest hole and a gentle repeated *piu piu piu* during nest relief and before feeding young. The soft, thin *peeuw* and *pweep* calls used in courtship can also be heard when a pair meets at the nest hole, or when they arrive with food for their nestlings. Very faint rising *wa* and *t-we* sounds are also made by adults by the nest. These can be hard to hear and are often only picked up by microphone when sound recording (Turner et al. 2022).

Anxiety call

Individuals also utter thin, rapid, strident, squeaky *we-we-we* notes when agitated or anxious. These can often be interspersed among 'laughing' advertising calls.

Gender differences in vocalisations

There are no obvious differences between the sexes in the various calls they make (at least to human ears in the field), although there may be some subtle distinctions in the advertising call. When spectrograms are studied, gender variations are sometimes revealed. Strong harmonics and monosyllabic fundamentals in versions by females can be seen to contrast with those of males, which may contain disyllabic fundamentals (Cramp

TABLE 1 Summary of adult call types, functions and phonetic descriptions.

Call description	Call function	Phonetic description
Advertising call (song/'yaffle'/'laughing' call)	Territorial proclamation, pair formation, contact	*kew, klu, kleu*
Regular call	Contact, arrival, departure, alarm	*tiew, kew, teuk*
Contact call	Contact with mate	*kuk, cook, kik*
Flight call	When moving within home range	*kjaek, kjeuk*
Flight alarm	When fleeing or flushed	*kju-kju-kjuk, kjack-kjack-kjack*
Alarm call	When disturbed or threatened	*kek-kek-kek, kik-kik-kik, kewk, kyak*
Threat call	During conflict	*kjaik*
Courtship call	Courtship	*peeuw, piu, pweep*
Calls by the nest	When by nest hole	*wa, t-we*
Anxiety call	When agitated or stressed	*we-we-we*

1985). Similarly, Turner et al. (2022) also found examples of the first note given by females being softer and lower pitched, with subsequent notes fairly evenly pitched, in contrast to those of males. If males and females do sing differently (and more study on this is needed), this would suggest that pair-bonding is involved.

Nestling sounds

Nestlings beg for food with harsh grating and whirring sounds. In the first week after hatching, they emit long series of continuous, low-pitched, rasping calls. These gradually develop into short, staccato, grating notes – not unlike the sound of cardboard being torn up – which they make when a parent

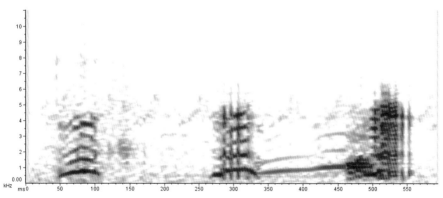

FIGURE 6.9 Nestling at approximately 2.5 weeks, showing a similar shape to an adult call in the first note, but sounding more like a 'squeaky rubber duck', plus two grating transitional notes. A second nestling gives a purer, rising note (coloured blue). Recording made in Lot, France, May 2011 (KT).

arrives. Sharp 'squeaky rubber duck' sounds are also made. As they grow and approach fledging, calls begin to resemble those of adults. All nestling calls tend to intensify and speed up when parents appear.

Fledgling calls

Fledged young utter *kejak* or *keyak* in series, similar to the calls of adult Black Woodpeckers, and sometimes a *peeyak* like a Jackdaw *Corvus monedula* (Gorman 2004). Juveniles are usually very vocal through July and into August, making soft clucks, or a double, frequently repeated *teu-tuo* call inflected on the first note, as they beg for food and maintain contact with their parents. As they start to become independent, fledged young make increasingly more adult-like calls.

Variations

Variation among the vocalisations of Green Woodpeckers differs more between individuals than regions. While the basic notes in vocalisations are the same or similar, the characteristics of rate, pitch, emphasis, progression and extent of repetition can all vary. Advertising calls are generally loudest and highest pitched at the beginning and decelerate towards the end, although there can be great variation between different birds and in the calls of individual birds, depending on the circumstances. An analysis of recorded advertising calls in 31 different territories across Europe (in England, France and Hungary) showed 80% falling in pitch, while 10% remained flat, and 10% rose (Turner et al. 2022). Sometimes these calls rise and then fall. Generally, the peak in frequency comes after the middle of the call and amplitude falls as the call progresses. The length tends to be greater during the breeding period, from March to May. In winter, isolated, short, piping advertising calls are occasionally made. Gentler and shorter versions can be exchanged between pairs, and between adults and young near to the nest. Any noticeable differences, however, are not linked to subspecies. For example, in Scandinavia birds are said to differ from those across continental Europe and Britain in terms of the tempo of their advertising call, which is slower, more decelerated and lower in pitch, despite the fact they are the same subspecies (Fauré 2018). It is curious that, although they are geographically separated, Green Woodpeckers in Britain do not seem to call differently from birds in continental Europe.

Drumming

Drumming by woodpeckers consists of a rapid, repetitive series of strikes with the bill on a substrate, usually a tree, and is distinct from the mechanical sounds produced during foraging or cavity excavation. This instrumental, non-vocal method of communication might be divided into two types: territorial drumming and soft drumming.

Territorial drumming

This is the best-known instrumental signal that many woodpecker species produce. It is typically loud, rapid, far-carrying, and used in home-range establishment, ownership and defence. As a long-distance signal it transmits this information to neighbours and potential rivals. It may also function as a way of attracting a mate and help to maintain the pair-bond (Gorman 2004).

Green Woodpeckers, however, probably do not use drumming as a territorial signal. In general, calls are more important modes of communication. For this species, in normal circumstances, drum-rolls match the soft drumming of other woodpeckers (Florentin et al. 2017). They are customarily feeble and lack any recurring structure in strike amplitude. Indeed, Glutz von Blotzheim and Bauer (1994) consider that instrumental signals have only negligible importance in communication for this species, but Blume (1996) states that the species can drum loudly. When Green Woodpeckers do drum, they typically produce a few, brief, barely audible, rumbling bouts during sessions of calling. In an analysis of 69 recordings involving 22 birds, the duration of rolls ranged between 327 and 2,324 milliseconds (average 1,282) with between 8 and 47 strikes per roll (average 26) (Turner et al. 2022). The rolls in this study typically decelerated, and some ended with tapping.

Soft drumming

In the same study (Turner et al. 2022), the most frequently heard examples of drumming were soft and were performed by partners, often in a loose duet, serving as an intimate method of communication, usually on the trunk close to a potential or active nesting cavity. The main functions of this kind of drumming are courtship, reinforcing the pair-bond and possibly nest-showing (Florentin et al. 2017, Florentin et al. 2020; Turner et al. 2022).

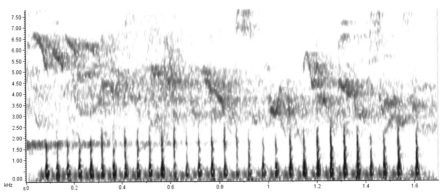

FIGURE 6.10 Drum-roll. This example shows a slowing strike rate. Recording made in Lot, France, March 2002 (KT).

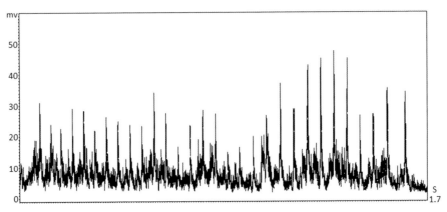

FIGURE 6.11 Pulse train analysis of a drum-roll (same recording as in Figure 6.10 above) showing the relative strike amplitude. This is a second feature of Green Woodpecker drum-rolls, in that they lack a regular pattern in the loudness of the strikes throughout a roll.

Tapping

Another form of instrumental signal is tapping, which is a brief and typically quiet sequence of strikes on a surface. It is always slower than drumming and does not form a clear roll as it exceeds the minimum of 90 milliseconds between strikes, which divides these two forms of instrumental communication (Turner and Gorman 2021). It may also be divided into two basic types: demonstrative tapping and nest relief tapping.

Demonstrative tapping

When tapping is done in discrete bursts of more or less equally spaced strikes it is termed demonstrative tapping. This acts as a signal, to advertise a potential nest site, to encourage a partner to participate in excavating a cavity or as nest relief tapping when birds change over during incubation or brooding. Such tapping may be combined with soft drumming, notably by the nest site during courtship. It is usually very gentle, difficult to hear and often only noticed when picked up by a microphone.

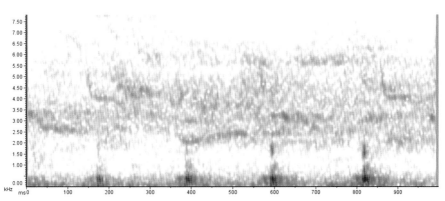

FIGURE 6.12 Demonstrative tapping. This example was given during a drumming duet. The four very soft taps were fairly evenly spaced, with average intervals of 215ms, while the force of each strike increased slightly. Recording made in Lot, France, March 2002 (KT).

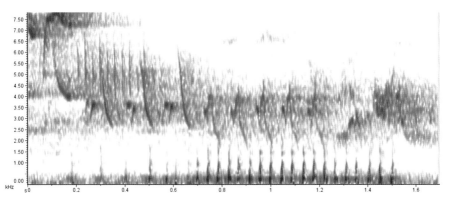

FIGURE 6.13 As mentioned above, drumming and tapping are sometimes combined. In this example, four taps preceded a drum-roll of 22 strikes. Recording made in Lot, France, March 2002 (KT).

Excitement pecking

Sporadic tapping sounds made near the nest have been described as 'excitement pecking' (Blume and Jung 1958). It is debatable, though, whether such sounds can be classed as ritualised instrumental signals in the way that demonstrative and nest relief tapping can be.

Use of artificial substrates

Typical woodpeckers drum on acoustically resonant snags, branches or trunks, maintaining regular, species-specific patterns which are either largely uniform or reduce as the rolls progress (Florentin et al. 2016). Green Woodpeckers do not appear to use favourite drumming posts or even select substrates on the basis of resonance. However, there are reports of Green Woodpeckers tapping and drumming on artificial surfaces, which suggests that some individuals may exceptionally do so. In May 1948 in Yorkshire, England, a Green Woodpecker was observed tapping on the circular metal caps at the top of three different electricity pylons (Dean 1949). The sound produced was described as like that of an automatic drill, audible at a distance of five to six field-lengths according to the direction of the wind. Intervals of from a few seconds to five minutes would elapse between bouts of tapping, during which the bird looked around as another called from a nearby woodland. This behaviour continued for several days, intermittently, from dawn to dusk. In March and April 2008 in Bedfordshire, England, another individual was watched over several weeks repeatedly tapping and drumming on the metal plates around the entrance holes of at least two nest boxes intended for songbirds (Kramer 2009). The sound produced was consequently amplified and could be heard over a hundred metres away. The woodpecker drummed four times in ten-second periods, followed by irregular tapping and then more drumming. Each bout lasted for about a second and usually consisted of 12–14 strikes, with eight recorded in one instance. The frequency slowed for the last two or three strikes. On most occasions the bird drummed three or four times, but also seven times in a minute followed by a further three times several minutes later. Perhaps significantly, three newly excavated holes were found in a tree about 8 metres from the box most frequently used for drumming.

In conclusion, Green Woodpeckers do employ instrumental methods of communication, but only anecdotal reports of them engaging in true territorial drumming exist. Rather, they produce soft drumming and demonstrative tapping, although these can be subtle and difficult to hear (Turner et al.

2022). The absence of loud, rhythmic, stable, territorial drumming in the Green Woodpecker (as in Middle Spotted Woodpecker and Eurasian Wryneck) constitutes a behavioural separation from most other woodpecker species within its range.

Chapter 7

Distribution, Population and Trends

The global distribution of the Green Woodpecker falls almost entirely within the Western Palearctic region, where it is resident and primarily sedentary. It occurs from Britain in the west to European Russia, the Caucasus and western and southern coasts of the Caspian Sea in the east, from Norway and Sweden in the north to Italy, the Balkans and Turkey in the south, and to the south-east there is an isolated population in north-east Iraq and south-west Iran. Despite much of the literature stating that there is a population of Green Woodpeckers in Turkmenistan, this is debatable. The species has only been observed in the extreme south-west of the country in the Köpet Dag Mountains and has not been recorded there since the early 1990s. It is possible that the birds found were vagrants from neighbouring Iran (Rustamov 2015).

Green Woodpeckers are absent as breeding birds from Finland and Baltic islands such as Bornholm and Gotland; Corsica, Sardinia, Sicily, Malta, Crete and Cyprus in the Mediterranean; Croatian islands in the Adriatic; Greek islands in the Aegean and Ionian; the Channel Islands; and Ireland, although vagrant individuals have been observed in some of these places (see Chapter 11, Movements and Flight). Although the species does not now breed in Sicily, it is thought to have been resident there until the early twentieth century, and there have been recent reports of individuals of this species on the island (La Mantia et al. 2015).

Elevation

Green Woodpeckers are generally found between sea-level and 2,000 m. However, the species can occasionally be found at higher elevations. It occurs at the timberline at around 2,450 m in the Meghri Range in southern Armenia

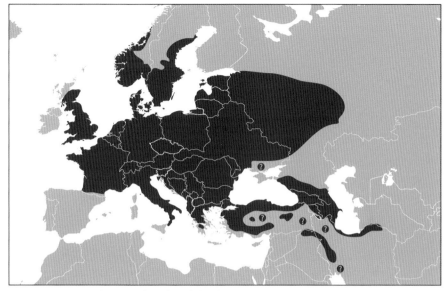

FIGURE 7.1 Map showing the global distribution (approximate) of Green Woodpecker.

and is said to breed at 3,000 m elsewhere in the Caucasus (Wilk 2020). In the Zagros Mountains of Iran, birds have been observed at around 2,000 m. In France it reaches around 2,100 m in the Alps and 2,000 m in the Pyrenees (Issa and Muller 2015). In Switzerland, some individuals occur above 2,000 m in Valais/Wallis Canton (Blume 1996). In Austria the species seldom occurs over 1,220 m (Weißmair and Pühringer 2015), but pairs have been found nesting at around 1,700 m and, in the southern Alps in Carinthia, they are occasionally recorded in the breeding season at around 2,000 m (Feldner et al. 2006). In Italy birds are sometimes seen at 2,300–2,400 m in the central and western Alps, and there is a breeding record from 2,100 m in the Aosta Valley (Brichetti and Fracasso 2020). In north-eastern Italy breeding birds are generally recorded at 1,100–1,200 m up to a maximum of 1,470 m. One study concluded that the higher altitudes are where there are traditionally farmed landscapes and a large proportion of broadleaved trees (Rassati 2005). In the Romanian Carpathians Green Woodpeckers occur at up to 1,300 m and in Bulgaria to 1,700 m (Iankov 2007). In the Czech Republic the highest elevation recorded for a breeding pair is 1,065 m in the Šumava Mountains (Šťastný et al. 2021), but in Slovakia and Poland the species is, for reasons that are unclear, seldom seen above 900 m (Matysek et al. 2020). Seemingly suitable montane pastures across this species' range are often not inhabited, most likely because of their long periods of deep snow cover which hinders foraging and feeding on the ground. Although Green Woodpeckers in Wales

FIGURE 7.2 High-elevation habitat at the timberline of oak and hornbeam forest at 2,254 m on the southern slope of the Meghri Range, Arevik NP, Armenia. Across their range Green Woodpeckers are rarely seen above this altitude (VA).

occur mainly in the lowlands, with few breeding records over 300 m, they are said to sometimes move higher to forage (Pritchard et al. 2021). This may simply be due to higher areas being well grazed and rich in ants but lacking suitable trees for nesting.

Population

Around 95% of the global population of Green Woodpeckers is considered to be in Europe and tentatively estimated at 587,000–1,050,000 breeding pairs. A very approximate estimate of 1,240,000–2,230,000 adult individuals has been made for the total global population (BirdLife 2022a).

Trends

The European distribution of this species has remained largely stable in recent decades, although some contraction has been noted in the Baltic States, Belarus and Ukraine (Wilk 2020). Its distribution outside Europe, in the easternmost and southernmost parts of its range, in Iraq and Iran, is less clear. A review of the population trends of 'widespread and common woodland birds' using data from an extensive European network of ornithologists for the period 1980–2003, classified the Green Woodpecker as a species that had experienced a moderate increase (Gregory et al. 2007; EBCC 2015). The most recent

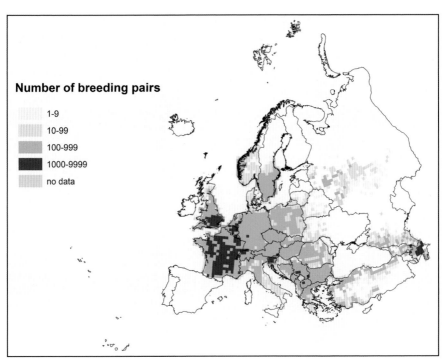

FIGURE 7.3 Map showing the breeding pair abundance estimates of Green Woodpecker in Europe according to the European Breeding Bird Atlas 2 (2020). The darker the area, the denser the projected breeding population.

atlas of European breeding birds states that numbers have increased in the west and declined in the north and north-east (Wilk 2020).

Detailed information from some countries is lacking, but where its population has been monitored a stable or increasing trend is typical. Several reasons have been put forward as being responsible for the rise in numbers in the west of its range, such as the maturing of some woodlands, favourable agri-environmental schemes and milder winters (Wilk 2020). On the other hand, in some northern regions where winters have begun to be milder, the species has declined. This is contradictory to projections based on climate change scenarios, and implies that several factors are influencing the species. Another positive trend that has been noted in several countries is this species' ability to colonise and thrive in urban areas (Battisti and Dodaro 2016; Kopij 2017; Fröhlich and Ciach 2020). Nonetheless, although it is clearly doing well in many places, residential and rural, some caution is advisable as in some areas a rise in observations may be linked to an increase in observers. As already mentioned, mild winters benefit Green Woodpeckers. Conversely, when temperatures are unusually low, they can suffer. These birds are sensitive

FIGURE 7.4 In many countries a trend of Green Woodpeckers moving into urbanised environments has been noted. As they are often rich in ants, even industrial sites with 'non-natural' habitats, such as this one, can be occupied. Drnholec, Czech Republic, August 2013 (TG).

to severe winters as deep snow and frozen ground affect their ability to forage for ants, and consequently local distribution and population densities are negatively affected (Nilsson et al. 1992; Glue and Südbeck 1997).

Throughout the twentieth century the species moved steadily northwards in Britain, although this range expansion was periodically checked by very cold winters (Glue and Boswell 1994). In the very north-east of England, in Northumberland, a rapid increase was said to have taken place from 1940 onwards, and by 1949 it was breeding in most of the deciduous woodlands of the region. Numbers fell again during the severe winter of 1946–7, but apparently soon recovered (Temperley 1951). In southern Scotland, sightings of Green Woodpeckers also began to increase from the late 1940s, although it is unclear whether this was due to more observers. A small influx in the south-east in the winter of 1950–1 heralded a colonisation and soon afterwards the first breeding pairs were documented in several border counties. Then, in the 1960s the south-west, east and central regions were colonised, and by the 1970s Green Woodpeckers had expanded into the north-east and were regarded as common in some parts of Scotland. By the early twenty-first century numbers had begun to fall back; today the species is still fairly widespread but localised and in small numbers, mostly south and east of the Great Glen. A few years after colonisation, declines in numbers of breeding pairs were noted in some areas. These were usually attributed to harder than usual winters, such as those that occurred in Lothian in 1979, Clyde in 1982, and Perth and Kinross in 1994 and 1997, which hindered ground feeding. Despite this, other chance events have also been suggested as contributing to the decline of several small and hence more vulnerable populations (Forrester and Andrews 2007).

In England, too, the hard winter of 1962–3 (the coldest recorded there since 1740) is said to have severely affected several bird species including Green Woodpeckers, with declines reported from many localities (Simms 1990). Birds in the south-west were seemingly less affected (Dobinson and Richards 1964). After another severe winter in Britain in 1981–2, local declines were again noted, some lasting several years (Glue 1993). Subsequently, a series of milder winters in the late 1980s probably then benefited the species, and more recently warmer winters have boosted the population (Smith 2007). With average winter temperatures now rising, the Green Woodpecker, as an essentially 'southerly' species, is projected to increase in population size and to colonise currently unoccupied woodlands in northern Scotland (Renwick et al. 2012).

FIGURE 7.5 A Green Woodpecker rests on a sapling and ruffles up its feathers to help keep warm. Kocsér, Hungary, New Year's Day 2011 (RP).

In Britain, and indeed everywhere, the distribution of the Green Woodpecker closely follows that of the ants on which it feeds (Kear 2003) and, as a warming climate is generally good for ants, the population trend is likely to continue to be positive. A recent study that evaluated the historical demography of the Green Woodpecker on the basis of DNA data and ecological-niche modelling concluded that historically Green Woodpeckers were sensitive to the effects of climate change (Perktas et al. 2015). Prior to the Last Glacial Maximum, around 20,000–25,000 years ago, a contraction of the population began, and the species became restricted to southern Europe, Anatolia and the Caucasus–Caspian region. Later, after the end of that glacial period around 14,500 years ago, it expanded northwards once again to occupy its present range.

Numbers and notes from selected countries

In the following summary, Green Woodpeckers are often described as being common and widespread, which are useful but rather relative terms. The natural densities of different bird species differ, so when Green Woodpeckers are described as being common or widespread in a country, that does not necessarily mean that they occur at the same densities as, for instance, many small passerine species, which are generally more abundant (Fuller 1982). Note that the figures below without a reference are drawn mainly from the European Environment Agency's Birds Directive 2013–2018 (2018) and/or from personal communications between the author and correspondents in the relevant countries.

Albania: Stable population of 1,220–1,830 pairs estimated.

Armenia: Population size and trends unknown, but probably more common than reported, and stable (Adamian and Klem 1999).

Austria: Common and widespread, absent only from intensively farmed areas and the highest Alpine zones. Increasing, with 17,000–28,000 pairs estimated (Dvorak 2019).

Azerbaijan: Fairly common and widespread, in both lowlands and uplands to the treeline. Population size and trends unknown (Patrikeev 2004).

Belarus: The rarest woodpecker species in the country, listed in the national Red Book, but widespread. Perhaps increasing but data lacking.

Belgium: Common and widespread, except in treeless polders and conifer forests of the Ardennes. Estimated 12,000 pairs nationwide. Increasing in Wallonia; in Flanders stable, perhaps increasing with 4,500–7,000 pairs estimated (Vermeersch et al. 2020).

Bulgaria: Common and widespread, but rare in upland conifer stands and absent from high elevations. Stable, with 10,000–30,000 pairs estimated (Iankov 2007).

Croatia: Fairly widespread. Generally more common in wooded hilly areas, such as in Istria, than in agricultural areas in the east. Absent from much of the barren coastal strip in Dalmatia and does not breed on islands. Trends unknown (Tutiš et al. 2013).

Czech Republic: Common and widespread. Estimated 9,000–18,000 pairs in the period 2001–3, then a slight increase to 10,000–20,000 over 2014–17. Now considered stable (Šťastný et al. 2021).

Denmark: Fluctuating trends around the country, but overall has declined, especially in northern Jutland. Some areas in the south recolonised. Probably under 400 pairs (DOF 2020).

Estonia: Very rare. Perhaps extirpated on the mainland. In 2017, 10–20 pairs were estimated to remain on Saaremaa island (Elts et al. 2019).

France: Common and widespread. Stable, perhaps a moderate increase, with 150,000–300,000 pairs estimated, which is the largest national population (Issa and Muller 2015).

Germany: Common and widespread. Stable, perhaps a moderate increase, with 42,000–76,000 pairs estimated. Short-term trend projected to be positive but long-term trend negative (Gedeon et al. 2014).

Great Britain: About 52,000 pairs are estimated and a long-term trend of increasing numbers predicted (Massimino et al. 2019). Widespread across southern England, patchier in the north. A strong increase first noted in the 1960s has continued, with a figure of 23% estimated for the period 1995–2019 (Harris et al. 2020). In Wales, the population remained roughly stable for most of the twentieth century, but a 29% decrease was noted 1995–2019, in contrast to neighbouring England. Some regions, however, such as Pembrokeshire in the late 1940s, saw a significant decline (Pritchard et al. 2021). There is a discernible east–west variance in the numbers in Wales. The south has always been a stronghold, with probably the largest population in Gwent, where 420–770 pairs were estimated 1998–2003. The situation in the east seems stable, but a decline in the west, particularly in Ceredigion, Meirionnydd, Pembrokeshire and Anglesey, has been documented in recent decades. By 2019 it had almost disappeared from these regions and had become rare in western Denbighshire. The total Welsh population was estimated at 3,000–3,750 pairs in 2018 (Pritchard et al. 2021). In Scotland, Green Woodpeckers have increased significantly since the first breeding pairs in the 1950s, but they are absent from much of west and north and patchy in the south-east. The general trend is one of colonisation, increases in breeding numbers, fluctuations and then local declines. The total Scottish population has been estimated at 600–900 pairs (Forrester and Andrews 2007).

Greece: Widespread in the north, becoming more uncommon and localised farther south. Breeds on the Peloponnese, but not on the islands. A stable population of 5,000–10,000 pairs is estimated.

Hungary: Common and widespread, occurring wherever there is suitable habitat. Stable until around 2008, then a strong increase noted. Most recent estimate is 15,000–17,000 pairs (Gorman et al. 2021).

Iran: National population size and trends are unknown, but said to be fairly common in some places, such as the Hyrcanian Forest in the north and the Zagros Mountains in the south-east, but uncommon in the north of the Azarbaijan region (Khaleghizadeh et al. 2017). Not officially protected.

Iraq: Only occurs in the mountainous north-east in Kurdistan where it is uncommon and localised, with 200–400 pairs estimated. Increasing in some areas but declining in others, in part due to illegal logging for timber, charcoal and fuel.

Italy: Widespread, though scarcer in the south and absent from mountain ranges and along the coasts of Calabria, east Basilicata and Apulia. Since the mid-1990s it has recovered in the Po Plain after a steep decline in the 1950s and 1960s. An estimated 60,000–120,000 pairs in the mid-2000s (Brichetti and Fracasso 2020).

Latvia: Very rare, with just 1–4 pairs. Has been in decline in range and numbers, for unknown reasons, since the 1980s (Ķerus et al. 2021).

Lithuania: Quite widespread, but more common in the east and south. One of the rarer woodpeckers in the country, with 500–800 pairs estimated (Brazaitis and Pėtelis 2010; Jusys et al. 2012).

Montenegro: Very limited data available as no systematic nationwide surveys have been done but said to be very common and widespread.

Netherlands: After a period of decline, has increased sharply since 1990. A downturn around 2010 thought to have been related to winters with deep snow. Estimated 8,000–9,500 pairs, mostly in the south and east (SOVON 2018).

Norway: Occurs mostly in lowlands, with highest numbers in the Oslo and Akershus areas. Seems to be in decline in Oppland. Estimated 3,000–6,500 pairs nationwide, but overall trend unclear (Shimmings and Øien 2015).

Poland: Widespread and moderately increasing but regarded as scarce overall. Estimated 15,000–26,000 pairs (Chodkiewicz et al. 2015).

Romania: Common and widespread. Stable, with 60,000–120,000 pairs estimated, which equates to one of the largest populations of the species (Petrovici et al. 2015).

Russia: Rare in the central part of European Russia, more common to the south in the Caucasus. Not included in the Red Data Book of all Russia but listed in those of some regions such as Bryansk Oblast, Ryazan Oblast, Moscow and St Petersburg (Prisyazhnyuk 2012). Trends largely unknown but thought to be in decline. Estimated 30,000–70,000 pairs in 2000–13 (Mischenko 2017) and a very wide possible 6,200–62,000 in 2020 (Kalyakin and Voltzit 2020).

Serbia: Widespread in both lowlands and uplands. Stable population of 9,000–13,000 pairs estimated (Puzović et al. 2015).

Slovakia: Fairly common and widespread, found wherever there is suitable habitat. Stable population of 1,200–2,000 pairs (Danko et el. 2002), with a strong increase of 110% estimated in the period 2005–9 (Slabeyová et al. 2009).

Slovenia: Common and widespread but absent from high mountains and areas of continuous forest. Stable, with 9,000–15,000 pairs estimated (Mihelič et al. 2019).

Sweden: Widespread and fairly common in the south; rarer in the north, where some populations have disappeared in recent decades. Steep decline from the mid-1970s until 2010. Now considered stable, with 18,000 pairs estimated in 2018 (Wirdheim 2020).

Switzerland: Stable trend in the period 2010–19, with 10,000–17,000 pairs estimated (Knaus et al. 2020).

Turkmenistan: Very rare. Formerly recorded in the very south-west in the Köpet Dag Mountains along the border with Iran, but no breeding has ever been confirmed. No observations since the early 1990s (Rustamov 2015).

Ukraine: Mainly occurs in the very west of the country. Has perhaps retracted in range. 700–1,000 pairs estimated (Wilk 2020).

Mortality

The number of Green Woodpeckers killed by wild predators, such as birds of prey and mammals, such as mustelids, is probably low (see Chapter 16, Relationships). Rather, Green Woodpeckers, like many birds that have colonised

FIGURE 7.6 Living in urban areas comes with its hazards. This unfortunate Green Woodpecker came to its end after flying into a window. Budapest, Hungary, April 2022 (GG).

FIGURE 7.7 Light snow cover does not present a problem for foraging Green Woodpeckers, but in hard winters with heavy snowfall they can suffer when ants move deep underground and out of reach. Kocsér, Hungary, July 2020 (RP).

settlements, are unfortunately often at the mercy of pet and feral cats. In fact, as Green Woodpeckers often forage on garden lawns, they can become easy targets for these non-wild predators. Human-related factors probably account for a significant number of deaths, too. Collisions with vehicles and windows are other dangers that birds living among people face (Cepák et al. 2008). Ironically, the oldest Green Woodpecker documented is a road-killed individual in England known from ringing data to have been just over 15 years old (Fransson et al. 2017). It should be noted that despite having evolved many shock-absorbing bodily features, woodpeckers are not 'super-avian', not indestructible. They certainly cannot sustain all impacts and are often fatally injured when they fly into windows and glass doors, for example.

Nevertheless, starvation owing to cold weather and heavy snow is most likely the main cause of Green Woodpecker mortality. Low winter temperatures, ground frost and deep snow negatively affect ants and therefore the woodpeckers. Consequently, northern populations and those at high elevations where densities are already low, such as in the Carpathians and Alps, are prone to declines during and after severe weather (Weißmair and Pühringer 2015). Locally, the effects of winter mortality may last for several years.

Chapter 8
Challenges and Conservation

Human-induced changes in the landscape of Europe have been negatively affecting wildlife, including woodpeckers, for centuries (Mikusiński and Angelstam 1997). As a species with a reliance on semi-natural grassland habitats as well as woodland, Green Woodpeckers may have fared well before the onset of the Industrial Revolution. In modern times, however, changes in land-use, particularly in farming and forestry, have accelerated and intensified as land has been cleared and replanted with non-indigenous trees to be harvested on shorter rotations, or ploughed to meet the demands of feeding a burgeoning human population. Thus, although they are generally regarded as doing well across their range, Green Woodpeckers have faced some challenges, as has much of the fauna with which they share their woodland and grassland habitats. The presence of these birds in a given area does not necessarily mean they are not under pressure. Intensification in agriculture and forestry has negatively affected the species. Farming without pesticides, herbicides and chemical fertilisers and with traditional livestock grazing is beneficial to ants and thus to Green Woodpeckers, but in many countries modern industrial agricultural methods have taken over. The use of chemicals, pasture 'improvement' or its conversion to arable land, and overly heavy ploughing and pollution all harm ants and hence have an impact on Green Woodpeckers (Nilsson et al. 1992).

In Britain, local declines in sheep farming result in fewer short-grass habitats for ants (Glue 1993; Glue and Südbeck 1997), as did the widespread reduction in rabbit numbers following the introduction of myxomatosis. In Wales differences in regional distribution are thought to be linked to land-use, particularly ploughing and re-seeding of permanent pasture, which affects ant availability. Reduced grazing due to fewer rabbits following a severe outbreak of disease during the 1990s, and cooler, wetter summers and mild, wet winters may also

FIGURE 8.1 When grasslands are ploughed up, ant colonies are often destroyed and consequently Green Woodpeckers disappear. Fejér County, Hungary, April 2022 (GG).

be contributing factors (Pritchard et al. 2021). Ultimately, the dependency that Green Woodpeckers have on ants is almost always at the root of the problems they face.

Orchards often provide ideal habitat for Green Woodpeckers (see Chapter 9, Habitats). Sadly, however, over the second half of the twentieth century cultivation methods in many old orchards across much of the range of the species were intensified. Newly planted or modernised orchards tend to be managed monocultures with few potential nesting trees and the overuse of herbicides and pesticides. In many places they were simply felled. In Slovenia, for example, more than 5,000 ha have been lost since 1990 (Vogrin 2011), and in Normandy, France 600,000 ha in 1950 had dwindled to 146,000 ha by 2000 (Collette 2008). In north-west Germany 75% of orchard meadows, perfect habitat for Green Woodpeckers, disappeared between 1979 and 2009 (Forejt and Syrbe 2019). A similar picture in England and Wales means that over half of the traditional orchards present in 1900 have now disappeared (National Trust 2022).

As Green Woodpeckers nest, roost and, to a certain extent, also forage in trees, they are directly affected – negatively and positively – by woodland and

FIGURE 8.2 This section of the trunk of a cherry tree housed a Green Woodpecker cavity but was felled in the nesting season. Zemplén Hills, Hungary, April 2015 (GG).

forestry management. The afforestation of land with conifers that occurred in many areas of Europe in the twentieth century may have benefited Green Woodpeckers in terms of food, as many ant species do well in such forests (Stenberg and Hogstad 1992). In north-east England, the widespread planting of coniferous trees, which intensified in the second half of the twentieth century, was advantageous for wood ants and consequently their abundance was said to have contributed to an increase in Green Woodpeckers in the region (Temperley 1951). Yet, in commercially managed forests, practices that include the removal of mature trees result in a dearth of potential nesting and roosting sites. Consequently, even when there is good feeding to be had, a lack of adjacent large trees where cavities can be made is detrimental.

Nonetheless, forestry can play its part in creating new habitats for a range of species, including woodpeckers (Quine and Humphrey 2010). Changes in habitat owing to human activity are, as is often the case, somewhat complex and Green Woodpeckers can be both winners and losers when landscapes are altered. Besides unfavourable farming and forestry, other human activities sometimes take their toll. For example, in the mountains of Iraqi Kurdistan, Green Woodpeckers, and indeed other woodland wildlife, are under increasing pressure from illegal logging for timber, charcoal and fuel, as well as fires caused by careless behaviour.

Conservation measures

There appear to be no comprehensive conservation activities, nationally nor internationally, that focus solely on Green Woodpeckers. Presumably this is because the species is often common and not regarded as a priority species by the IUCN. Ultimately, overall nature-friendly forestry and farming and general conservation projects that conserve woodlands and semi-natural grasslands are beneficial for this woodpecker (Alder and Marsden 2010). Locally, the preservation of mature trees, and standing deadwood, can also help the species. Green Woodpeckers can benefit from careful woodland management in floodplains, where coppicing and selective felling create open habitats with grassy clearings (Spitznagel 1990). In some areas, they are affected by both the loss of suitable nesting trees and the degradation of foraging areas by intensive management and/or the use of pesticides. However, changes in farming and forestry practices to address biodiversity loss and climate change may well give cause for optimism (European Commission 2020).

Green Woodpeckers do not often take to breeding in artificial nest boxes; therefore, erecting these does not adequately address any lack of suitable cavity trees. Rather, in areas with a dearth of old trees where cavities can be made, nest boxes are sometimes used as roosting sites (see Chapters 10 and 13). Parks, gardens and cemeteries can be important habitats for woodpeckers in urban areas if they contain some deadwood (branches, snags, stumps, logs) and large old trees. Deadwood resources, which host high numbers of invertebrates, especially beetle larvae, should always be included in residential landscape planning and green space management, for all woodpeckers and indeed for many other species (Fröhlich and Ciach 2020). In order to ensure that ants are abundant, garden lawns, sports fields and farmland should be managed without resorting to pesticides.

FIGURE 8.3 An adult female Green Woodpecker using a broken snag as a lookout post. Though deadwood such as this is seemingly not as important for the species as it is for some other woodpeckers, it is nevertheless a valuable resource and should always be left in place. Bulgaria, July 2020 (DG).

Conflicts with humans

It is probably safe to say that Green Woodpeckers are liked, but at times their behaviour can test people's patience. For instance, some peck into honeybee hives to prey on the insects within, which, of course, does not endear them to apiculturists. The birds can easily break through the thin wood and polystyrene of artificial hives. In addition to the many bees consumed, the holes made can lead to the loss of the whole colony as they are too big for the bees to repair and the cold air that enters is fatal. To address this problem, bee-keepers will put frames of chicken wire, garden mesh or netting around hives to prevent woodpeckers getting at their bees.

Green Woodpeckers will also make holes in wooden utility poles (which irritates energy and telecom companies), in wooden buildings and in polystyrene wall insulation (which can infuriate builders and homeowners). In most countries this species is protected by law, so culling problem birds is not an option. As with damage to beehives, prevention is better than cure, and a range of deterrents are available, such as 'scare-eyes'; owl, falcon and hawk dummies; flash-tape and reflective discs, although it has to be said that these do not always work. Another method used to discourage them is to paint surfaces with non-toxic ammonium sulphate compound.

FIGURE 8.4 A Green Woodpecker peeps out of a roosting hole that it excavated in the wall of a hotel. Olomouc, Czech Republic. July 2021 (TG).

Umbrella species

An umbrella species is one that has a relatively large home range and distinctive habitat requirements, both of which overlap with other organisms. Species are designated as being 'umbrella' by ecologists and conservationists because protecting them ultimately protects the many other species that make up the ecological community they inhabit. Woodpeckers in general can be considered umbrella species because conserving their woodland and forest habitats (and, in the case of Green Woodpecker, also grasslands) consequently benefits innumerable other species. Other birds, of course, but also many mammals, amphibians, reptiles, invertebrates, fungi and plants can thrive under the Green Woodpecker's umbrella; although their habitat requirements may initially seem rather different, they are all interconnected.

Indicator species

In ecology a species that demonstrates the condition and quality of a habitat or ecosystem by virtue of its presence there is termed an indicator species. As a group, woodpeckers are considered to be excellent indicators of biodiversity in wooded habitats (Mikusiński and Angelstam 1997; Lõhmus et al. 2016). As most, including the Green Woodpecker, are resident and often sedentary, they are more reliable as indicator species than migratory birds, whose populations are affected not only by conditions in their breeding areas, but also by those encountered when they are on migration and in the wintering quarters. Green Woodpeckers need mature trees for breeding and roosting cavities and agriculturally 'unimproved' grasslands for feeding, and if old trees are removed or grasslands 'improved' or heavily ploughed, the species will not occur in an area. Many other species also depend upon the same ecological requirements as Green Woodpeckers for their existence. Therefore, it seems reasonable to declare that Green Woodpeckers are important natural indictors of the ecological health of combined woodland–grassland habitats.

The Green Woodpecker is one of the bird species that is advocated as an indicator of High Nature Value (HNV) farmland in Europe. HNV is a concept that recognises and promotes low-intensity farming systems which are valuable for biodiversity and the environment. Green Woodpeckers are often found where there are semi-natural pastures, meadows, orchards, tall hedges and copses, all of which are recognised as key features in HNV farmland. In a study of a radio-tracked pair of Green Woodpeckers in Dorset, England the birds were found to select grassland feeding sites that had the most species of wild plants (Alder and Marsden 2010).

FIGURE 8.5 Unimproved grazed or mown grasslands (on the right) provide good foraging habitat, unlike heavily ploughed land (on the left). Fejér County, Hungary, April 2022 (GG).

Flagship species

A 'flagship species' is one that is used to raise the awareness of conservation issues and generate support and funding for campaigns. Most flagship species are chosen because they are iconic, charismatic, popular and easily recognised by the general public. Classic examples, known worldwide, are the Giant Panda and Bengal Tiger, but not all flagship species are as big, impressive and familiar as these. On a non-global level, smaller animals are often chosen. Woodpeckers in general are charismatic and popular birds and are regarded as keystone species (see Chapter 16). As such they can be ideal local 'flagships', acting as avian ambassadors for conservation organisations or projects. For

FIGURE 8.6
The Green Woodpecker was 'Bird of the Year' in Hungary in 2022. A range of items – posters, calendars, bookmarks and this sticker – were produced to highlight and promote the bird and its habitats (GG).

example, in Hungary one species is picked every year to be 'Bird of the Year'. In 2022 the Green Woodpecker took on that flagship role and proved to be hugely popular, especially with children.

Conservation status

The IUCN focuses on nature conservation and the sustainable use of natural resources. The organisation maintains a Red List, which classifies species according to their global extinction risk. The highest level of threat is termed 'Critically Endangered' and the lowest 'Least Concern'. Green Woodpeckers are distributed over a very wide area and, although there have been declines locally, often considered to be a result of grassland and woodland degradation, the overall trends seem positive, and thus the species is not deemed in danger. In Britain, a traffic-light system (Red, Amber, Green) is used to assess the level of conservation concern for birds. In a review in 2015, the Green Woodpecker was moved from Amber to Green, the lowest level of concern, after having recovered from an apparent decline (Eaton et al. 2015). In the most recent assessment, it remains in the Green category (Stanbury et al. 2021). At a national level, the Green Woodpecker is a legally protected species in most of the countries where it occurs.

Chapter 9

Habitats

Green Woodpeckers are typically described as woodland birds, but this is a simplification as they also habitually use non-wooded habitats when feeding. The species inhabits a variety of wooded-grassland environments across its global range, which is not unexpected given that it is distributed over three eco-climatic regions: the Mediterranean, temperate and boreal. Wherever they occur, Green Woodpeckers require two fundamental habitats to fully satisfy their ecological needs: open mature broadleaved woodland and short-grazed or mown grassland.

FIGURE 9.1 Grasslands with high numbers of terrestrial ants are essential foraging habitats for Green Woodpeckers. Zagreb, Croatia, April 2016 (MS).

Wooded habitats

Broadleaved and broadleaved–conifer woodlands and forests are favoured over homogeneous coniferous stands. A study in Britain found nests mainly in large tracts of mature deciduous or mixed broadleaved–coniferous woodland, but only occasionally in pure conifer stands (Glue and Boswell 1994). Even in the very north of their range where coniferous forests dominate, such as in Norway, Green Woodpeckers are more frequently encountered in mixed forests with a large proportion of deciduous trees (Stenberg 1994). Although softwood trees are generally favoured over hardwoods for nesting sites, this does not mean that hardwood woodlands are avoided. In fact, a study in the floodplain forests of the Donau-Auen National Park in Austria found that Green Woodpeckers preferred hardwood stands to areas with softwoods (Riemer et al. 2010). Poplars and willows were apparently avoided, but this was attributed to the wetter ground conditions where these trees grew. Quite simply, drier areas where hardwoods dominated hosted more ants. Parkland, semi-natural woodland with clear-cut areas, young open plantations with adjacent grassy rides and farmland with copses and belts of trees are also used. Old-growth forests, with snags and other kinds of deadwood, are not as important for Green Woodpeckers as they are for most of the other woodpeckers with which they are sympatric. However, a study in Britain concluded that pairs required mature timber for nesting (Glue and Boswell 1994). On the other hand, a later study in southern England found that Green Woodpeckers showed no preference at all for dead trees as nesting sites (Smith 2007).

Grassland habitats

The Green Woodpecker is (along with the Iberian Green Woodpecker) the most terrestrial member of the family in Europe, selecting open grassland habitats where ants are common and easy to obtain (Gorman 2004; Alder and Marsden 2010). Heaths, paddocks, pastures, meadows, and grassy coastal and inland dunes are all used. Areas that are not treated with herbicides, insecticides or pesticides, and thus where insect prey thrives, can be vital.

The structure of these grasslands is important, with taller, denser vegetation generally avoided by feeding birds in favour of short swards where ant prey is common and accessible (Alder and Marsden 2010). Drier grasslands with a sunny aspect are preferred over shady, damp and boggy ones because they generally harbour high numbers of ants (Glue and Boswell 1994; Glutz von Blotzheim and Bauer 1994; Cramp 1985; Blume 1996). The well-documented

FIGURE 9.2 Grasslands grazed with livestock such as this paddock with horses are ideal foraging habitat. Fejér County, Hungary, April 2022 (GG).

FIGURE 9.3 An adult male Green Woodpecker stands on a rock in a grassy field to scan around for signs of food. Kocsér, Hungary, June 2020 (RP).

range expansion through Scotland since the 1950s was slower in the west, and it continues to occur rather patchily there (Forrester and Andrews 2007), which may be due to the wetter Atlantic-influenced ground conditions in that region. In floodplain and riverine woodlands, such as along the Rhine and Danube, densities are usually highest in more open landscapes where there are grassy dykes and some old broadleaved trees (Spitznagel 1990; Riemer et al. 2010).

New grasslands

In Britain, as in many parts of continental Europe, this species has adapted to forage on farmland, notably permanent grazed pastures, and lawns in residential areas (Alder and Marsden 2010). These 'new' grasslands have increased considerably since Britain's wildwood was cleared. Nevertheless, it is important to note that much of Britain's – and indeed Europe's – agricultural land is unsuitable for Green Woodpeckers. The species cannot survive where intensive farming practices, such as the use of chemicals that destroy ant colonies, are employed (Glue 1993). The ploughing up of old pastures can also result in pairs abandoning an area, but if the improved pasture reverts to its former condition, they can return. Conversely, land abandonment causes alterations in the structure of grasslands which then affects prey availability (Muschketat and Raqué 1993). Green Woodpeckers are sensitive to changes in stability in these habitats resulting from successional change, particularly in the early stages after a shift in land-use practice (Nilsson et al. 1992).

Ecotones

Woodlands in the period after the last glaciation would have reached their optimum around 8,000–10,000 years ago. Woodlands across prehistorical Europe were likely to have been highly dynamic and varied rather than uniform. Most were probably a mosaic of closed-canopy and more open woodland, scrub and grasslands, which would have been subject to many influences from large grazing herbivores and natural storm and flooding as well as fire events. Where these happened, grazing and browsing animals would take advantage of the young regrowth and it is likely that many areas were kept open for long periods, providing suitable foraging habitats for open-habitat species including Green Woodpecker. The impact of humans was likely to have become significant in the Mesolithic from 8,000 years ago and, like other species, people moved into landscapes as they became more hospitable. In other words, we as a species have been influencing the landscape for many

FIGURE 9.4 Edge habitat. Broadleaved woodland that has trees large enough for nesting coinciding with grassy parkland with healthy numbers of ants and other insect prey represents a perfect combination for Green Woodpeckers. Buda Hills, Hungary (GG).

millennia (Bradshaw et al. 2003). Green Woodpeckers would have exploited areas of open habitat such as on limestone grasslands and heathlands where soils were poor and where ants were found both naturally and in places people had cleared vegetation to encourage game animals which they hunted. Today, in all regions, Green Woodpeckers seldom enter dense, closed forests, though they will visit clearings within them. Indeed, they are often found in transitional habitats: the ecotone between wooded areas where they nest and grassy areas where they forage on the ground. Hence, the Green Woodpecker might be regarded as an 'edge species' (Mikusiński 1997).

Uplands

Though the bulk of the global Green Woodpecker population occurs at low elevations, and the species is generally regarded as a 'lowland bird', pairs can be found in uplands. When they occur in mountainous regions, like the Alps, Carpathians and Balkan ranges, Green Woodpeckers generally prefer warmer, south-facing woodlands where there are open grassland habitats such as grazing pastures, meadows, clearings and windfall areas (Glutz von Blotzheim and Bauer 1994; Weißmair and Pühringer 2015). In the Zagros Mountains of Iran, they occur in open oak forest (Kaboli et al. 2016).

FIGURE 9.5 Traditionally grazed pastures within wooded uplands can be suitable foraging and nesting habitat for Green Woodpeckers. Lilienfeld, Lower Austria (TH).

FIGURE 9.6 Green Woodpeckers often do well in traditionally managed orchards with open, unimproved grassland. In such places the combination of trees for nesting and plentiful ant prey is ideal. Eschenau, Lower Austria (TH).

Orchards and vineyards

Orchards, groves and vineyards can be excellent habitat for Green Woodpeckers, as indeed they are for some other picids such as Syrian Woodpecker *Dendrocopos syriacus* and Wryneck *Jynx torquilla* (Gorman 2022). Indeed, in southerly regions they are often among the most important habitats used (Ruge 2017). Traditionally managed orchards are ecologically akin to natural wooded-grassland habitats and their avifauna is much richer than it is in intensively managed orchards (Kajtoch 2017; Chmielewski 2019; Barker 2021). Non-intensively managed vineyards with adjacent woodland, or at least some larger, old trees where cavities can be made, are also frequently visited for foraging.

Urban habitats

Ultimately, native open broadleaved woodlands provide the best habitat for this species. Yet it has become increasingly synanthropic, spreading into areas where human habitats have modified or replaced natural ones. Today, the Green Woodpecker often inhabits parks, gardens, arboretums, cemeteries,

FIGURE 9.7 Inner-city parks with grassy areas for foraging and some large trees for nesting sites, are often inhabited, even when human activity is commonplace. Nevertheless, Green Woodpeckers never become 'tame' and readily flush when approached. Budapest, Hungary (GG).

FIGURE 9.8 A female Green Woodpecker clings to a tree in a cemetery. In many countries these birds have moved into such wooded urban habitats. Olomouc, Czech Republic, June 2018 (TG).

golf courses, sports fields and even inner-city areas. Across its range only the Syrian Woodpecker and the Great Spotted Woodpecker *Dendrocopos major* have taken to such environments to a similar degree.

Former industrial and military sites, which are typically mosaics of trees, bushes, grassy patches and concrete pavement, where some ant species thrive, are also often colonised. In Wales, for instance, pairs have settled in old colliery sites, which in recent times have reverted to nature (Pritchard et al. 2021). Where Green Woodpeckers have moved into settlements, the two key factors for their success are the same as in semi-natural and rural areas: the presence of ground-dwelling ants and other invertebrates, and some mature trees where cavities can be made.

Sharing habitats

The typical Green Woodpecker environments mentioned are not exclusive to the species. Within their range, ten other woodpecker species can potentially share these habitats. Depending on location, Iberian Green, Grey-headed, Black, Great Spotted, Syrian, White-backed *Dendrocopos leucotos*, Middle

FIGURE 9.9 Green Woodpeckers share habitats with several other picid species including, as here, Middle Spotted Woodpecker in a park in Sofia, Bulgaria. January 2022 (MS).

Spotted *Leiopicus medius*, Lesser Spotted *Dryobates minor* and Eurasian Three-toed Woodpeckers, as well as the Wryneck, may all co-occur alongside Green Woodpeckers. White-backed and Eurasian Three-toed Woodpeckers are, however, seldom found in the same areas as Green Woodpeckers, owing to the more arboreal foraging requirements of those two species. Over much of the Green Woodpecker's continental range, Grey-headed Woodpeckers are often found sharing its habitat (Raqué and Ruge 1999). For more on this see Chapter 16, Relationships.

Trees

Green Woodpeckers do not have a close ecological connection, either for foraging or nesting, to any particular tree species. They are only associated with specific trees at a local level. Both native and introduced trees are used for nesting, provided the bole is large enough to house a cavity and there is an area of softer wood, often due to fungal decay, to make excavation easier (Gorman 2020c). Trees stricken with heart rot are frequently chosen by woodpeckers for the locations of their cavities and the Green is no exception. Pairs will nest in both live, dying and dead trees, but dying or living ones with soft areas resulting from fungal decay or insect infestation, or damage from lightning strikes, strong winds or frost, offer better opportunities for excavation than those that are healthy, sound and hard. Moreover, the openness of wooded areas and availability of ant prey is more important than the species of tree (Spitznagel 1990; Rolstad et al. 2000; Riemer et al. 2010). Nevertheless, broad-leaved trees are generally preferred over coniferous ones (Glue and Boswell 1994). Obviously, as Green Woodpeckers occur in diverse eco-climatic zones and in habitats from high mountains to seacoasts, the type and species of trees they use varies considerably. In simple terms, we might say that once they have discovered an area with plenty of food resources, they can only use the trees that are locally available. The following are some examples.

Thirty-three nesting cavities in Hungary, documented between 2006 and 2020, were located in 13 different tree species (Gorman 2020c). All were broadleaved, as follows, with the number of times used in brackets: Common Alder *Alnus glutinosa* (1), Ash *Fraxinus excelsior* (5), Beech *Fagus sylvatica* (3), Field Elm *Ulmus minor* (2), Hornbeam *Carpinus betulus* (2), Horse Chestnut *Aesculus hippocastanum* (1), oaks *Quercus* spp. (5), poplars *Populus* spp. (4), planes *Platanus* spp. (1), Small-leaved Lime *Tilia cordata* (3), Sycamore *Acer pseudoplatanus* (1), Walnut *Juglans regia* (2) and willows *Salix* spp. (3). In the extreme south-east of their range in the Zagros Mountains and the Golestan

National Park in Iran, Green Woodpeckers mainly inhabit oak forests (Kaboli et al. 2016; Varasteh Moradi et al. 2018). In Azerbaijan, they use old walnut and chestnut trees (Patrikeev 2004). In the Köyceğiz area in Turkey predominately Anatolian Sweetgums *Liquidambar orientalis* are inhabited (Ürker and Benzeyen 2020). A study in Britain found a wide spectrum of larger broadleaved, rarely coniferous, trees were used for nesting sites, mainly oaks and Ash, but also occasionally birches *Betula*, Beech, Elm, Aspen *Populus tremula*, Hawthorn *Crataegus monogyna*, Common Hazel *Corylus avellana*, White Poplar *Populus alba*, Hornbeam and Walnut (Glue and Boswell 1994). Despite the forests of Scandinavia being mainly coniferous, the broadleaved Aspen is favoured by Green Woodpeckers which depend upon it for nesting, with the highest numbers in areas with this tree (Stenberg and Hogstad 1992; Rolstad et al. 2000). Similarly, in Scotland, although they often forage in stands of pine *Pinus* spp. Green Woodpeckers tend to breed in broadleaved trees, especially birch (Forrester and Andrews 2007).

The evidence seems to suggest that although at the northern extreme of their distribution and at high elevations Green Woodpeckers may be compelled to use conifers, this is not ideal. A study in Sweden determined that a decrease in broadleaved woodland was one of the reasons that the species had declined there (Nilsson et al. 1992).

Chapter 10

Behaviour

Green Woodpeckers, like all woodpeckers, are diurnal. They are typically active soon after dawn and go to roost before dusk. Day length and light intensity seem to play a key role in determining activity and behaviour. Outside the breeding season, most of their day is spent foraging and flying between good feeding sites. In the middle of the day, they often indulge in comfort behaviour (see below) such as preening or simply rest in trees. In the breeding period their activities and actions change dramatically (see Chapter 12, Breeding). Interactions, including courtship and agonistic interactions with other Green Woodpeckers and other bird species, become more frequent in early spring as the hours of daylight increase (Keicher 2007). Similarly, interactions become fewer in late summer and autumn once offspring have finally dispersed.

Sociability

As is usual for most species in the *Picus* genus, Green Woodpeckers are neither gregarious nor overly social. They are generally solitary, often appearing to have a low tolerance for their own kind. They do not gather in flocks, although they can be seen in family parties for some weeks after fledging. Neither do they forage cooperatively and even established pairs rarely feed together. The exception is when it is time to breed, but even then the close bond that is apparent in many other monogamous birds is not always obvious. The sexes collaborate in raising their brood, sharing incubation, feeding and nest sanitation, but they tend to work in parallel. Nevertheless, outside the breeding season a pair may roost in holes close to each other, but seemingly never together, and they will call and respond to one another, thus maintaining a loose relationship (Blume 1996).

FIGURES 10.1a, 10.1b, 10.1c An adult male Green Woodpecker interacts with a recently fledged juvenile male. The parent bird's rather aggressive behaviour is probably an attempt to force the young bird to disperse from the natal area and begin fending for itself. Padova, Italy, July 2022 (CC).

Dominance

Green Woodpeckers protect important resources, such as trees with roosting cavities and productive ant colonies, from intruders of their own kind, regardless of sex. In periods of ant abundance, such as when these insects swarm in warm weather, several individual woodpeckers may attempt to congregate but dominant individuals usually restrict this. Most disputes between two birds are, however, same-sex confrontations. When the two sexes do come into conflict, males are invariably dominant. A male Green Woodpecker may point his bill towards his mate and the bills may even touch, but rather than a display of dominance this may be a form of symbolic feeding. Fledged young are tolerated at feeding sites and sometimes fed, one or two following each parent, until late summer or early autumn before they become independent. After that, young birds are rarely accepted by their parents if they return to feeding sites that they previously, and even very recently, shared.

Antagonistic displays

Green Woodpeckers seldom physically defend their home ranges, which do not have clearly defined borders, from each other. It is likely that calling mitigates the need to do this. Rather, they confront each other at more precise locations such as roosting cavities and feeding sites within a larger vague territory. Any confrontations that do occur are rarely physical, although rivals may lunge at each other. Potentially injurious contact is averted by posturing displays that involve body-swaying, head-bobbing, wing-spreading and -flicking, crown-raising and ruffling of the body feathers. Aerial pursuits often end with rivals pursuing each other around a tree trunk in a spiralling chase, then stopping, pointing, swaying, and making figure-of-eight patterns in the air with the bill. When engaging on the ground, Green Woodpeckers will point their bills skywards and sometimes whip their long tongues in and out. One or both birds will occasionally pause and feign disinterest, staying immobile on the ground or on opposite sides of a tree trunk for some minutes before initiating the confrontation again (Blume 1996). Altercations can also be suspended, with one or both individuals mock preening. Antagonistic displays are interspersed with periods of calling and silence.

Courtship displays

As with most woodpeckers, the courtship behaviour of Green Woodpeckers includes many coordinated and strikingly ritualised postures and movements that can appear hostile, recalling their antagonistic ones. Courtship displays usually take place on the ground, but occasionally on a bough, near the future nesting cavity (see Chapter 12, Breeding). Displays vary from pair to pair, but some elements are fairly standard, such as when two birds face each other and rhythmically sway and jerk their head from side to side, with their necks stretched out and bills pointed upwards. This display may be done before or after courtship-feeding and copulation. Variations of this basic display include the pair swaying in sync in the same directions or swaying in opposite directions. They may bow in unison, sometimes one or both birds stretching their head right back and pointing the bill skywards. They may also move their bills in a flattened U-shaped arc dipping in the centre of the swing (Snow and Manning 1954). Both birds may suddenly freeze, remaining motionless

FIGURE 10.2 Faced with a perceived threat, Green Woodpeckers often move to the opposite side of a tree trunk from where they cautiously assess the situation. Budapest, Hungary, January 2009 (SK).

for a minute or so, and then move around, approaching each other and/or retreating. Displaying birds usually stand more than a bill-length apart, but occasionally close enough to touch bill tips. They may display in silence or utter soft calls. Courtship-feeding involves the female pecking at the male's bill and taking any food he then offers. Despite the pair-bond being relatively weak outside the breeding period, apparent courtship-feeding in August has been observed (Hann 1951).

Vigilance

Considering that they often occur in urban habitats where they come into daily contact with people, in places such as gardens, parks and cemeteries, Green Woodpeckers are remarkably wary. They are typically hard to approach, easily disturbed and readily flush. They are also notably cautious when going to and leaving their overnight roosts. When confronted with a human intruder, they will often simply flee or move to the opposite side of a tree trunk, peeping

FIGURE 10.3 Sensing danger, a male Green Woodpecker watches intently from the cover of a tree. Toulon, France, March 2020 (JMB).

out from time to time to assess the threat. On the other hand, when caught and handled by bird ringers, they are usually feisty, often making raucous alarm calls, and given the chance will aggressively peck the handler. When foraging on the ground, Green Woodpeckers frequently look up – seemingly vigilant for aerial predators such as falcons and hawks. In response to non-avian threats at the nest, such as from squirrels and martens, Green Woodpeckers are usually submissive unless their nestlings are at risk. There is, however, some difference in how individuals react, some being bolder and more assertive than others.

Passivity

When disturbed while in a tree, Green Woodpeckers often do not flee at once, but move behind the trunk and hide, peeping out from time to time to monitor the situation. They may also simply freeze, remaining motionless on a branch for several minutes, usually with their head and bill pointed upwards. Though this is probably part of a passive defensive strategy, they evidently also sit still at other times, as if sleeping (Cramp 1985). When in a nesting cavity, most adults sit tight, though there is individual variation, with certain birds immediately flushing when sensing potential danger.

Displacement behaviour

Birds, along with other animals and humans, sometimes react to stressful situations by pretending to ignore them. Rather than face a threat or continue a dispute, they respond by behaving in a manner that seems unconnected to the situation. When in trees, woodpeckers may start pecking at a branch to feign disinterest, although this is futile when faced with a determined intruder. A common displacement response of Green Woodpeckers is mock preening. This behaviour is generally only performed when perched in trees, as when confronted on the ground Green Woodpeckers usually just flee, perhaps because they feel too exposed and vulnerable there. All in all, responses depend upon the situation and when real danger threatens Green Woodpeckers often simply fly into cover. They do not indulge in paratrepsis (feigning injury as an anti-predator distraction), which is a more extreme form of distraction than displacement behaviour.

Comfort behaviour

Green Woodpeckers, like all birds, practise comfort behaviour, a term that describes actions related to body care and which presumably enhance physical comfort. Preening, scratching, stretching, basking in the sun, anting, dusting, and bathing in water are all forms of comfort behaviour. As they are often on the ground, Green Woodpeckers probably engage in dusting and anting more often than their more arboreal relatives but, as they are wary, these activities are carried out discreetly and so are not often observed.

Preening

Preening is an essential part of feather care and maintenance. It is essentially a cleaning process where dirt and ectoparasites are removed and lubricating oil applied. Preening can be done at any time but frequently happens just after feeding or excavating work, presumably because the feathers are often soiled during those activities. Green Woodpeckers typically preen for several brief periods during the day rather than in one prolonged session, and this is habitually performed in association with the other comfort behaviours mentioned above, either before, after or interspersed with them. A typical session of preening involves body-shaking, feather-ruffling and the more precise drawing of individual feathers through the bill. To reach some areas of plumage, birds stretch and contort themselves: the neck twisted so the bill can reach the mantle and the wings lifted in turn. Green Woodpeckers do not allopreen; that is, they do not preen each other.

Sunning

Basking in the sun, or perhaps simply sunbathing, typically takes place on exposed trunks, boughs, snags or utility poles, and sometimes on the ground. It is often interspersed with preening. When sunning, Green Woodpeckers usually face the sun, periodically raising the crown feathers, opening the bill, ruffling the feathers, fanning out the tail, lifting the wings and stretching (Glutz von Blotzheim and Bauer 1994). Individuals sometimes appear to doze for some minutes when sunning. It is unclear why birds indulge in this behaviour. Perhaps it is simply for comfort and pleasure, or to feel the warmth? Two observations in Germany provide intriguing insights into this comfort behaviour. The first, in October 1999 at 5 p.m. when the temperature was 5°C, involved a male clinging to a trunk tree some 90 cm above the ground. The second, in February 2000 at 8:10 a.m. when the temperature was just 2°C, involved a juvenile, also on a tree trunk, around 120 cm above the ground, which

FIGURE 10.4 A female Green Woodpecker preening. Kocsér, Hungary, January 2010 (RP).

faced towards the rising sun. On both occasions the birds appeared to be slumbering, lethargic and thus, unusually, allowing the observers to approach to within 2 metres (Reichholf 2001). Alternatively, sunning probably serves a function related to feather health, as birds often preen during or immediately after sunning. This behaviour probably also helps produce vitamin D, an important component in preening oil, and discourages or displaces parasites such as ticks and mites.

Bathing

Green Woodpeckers will bathe in both natural and garden ponds, and in puddles and pools. They are not often observed doing so as they are especially wary at such times, no doubt aware that they are more vulnerable to predators in these moments. They will step into water but appear agitated and uneasy, regularly stopping to glance up and around. Consequently, close-up images of Green Woodpeckers bathing have in recent times usually been obtained at purpose-built artificial water-features placed in front of photographic hides and by camera traps.

Dusting

Dust bathing is a form of comfort behaviour that is probably more common in Green Woodpeckers than in most other woodpeckers as they are habitually on the ground. Dust baths are usually temporary sites, located on dirt tracks and dry bare ground (Glutz von Blotzheim and Bauer 1994). Dusting probably helps eradicate parasites and so, as with sunning, is likely done to maintain feather health. It is also sometimes combined with anting.

Anting

Anting is a form of behaviour where birds allow, or encourage, ants to eject the acidic chemicals that they release when disturbed onto their feathers. Though not often witnessed, anting is probably common in many birds, including woodpeckers, but its exact function is still not fully understood. Explanations include the ridding of ectoparasites, feather care, lessening skin irritation during moult and sensory self-stimulation. Another theory (which has generally not drawn much support) is that anting is food preparation, where ants are provoked to squirt their acid before being eaten in a more palatable form (Judson and Bennett 1992). However, none of these hypotheses has been convincingly proven (Morozov 2015).

There are two kinds of anting: passive and active. Passive anting involves birds simply squatting among the insects, letting them run over the body

FIGURE 10.5 A juvenile Green Woodpecker indulging in a session of passive anting and stretching. Norfolk, England, August 2011 (NB).

FIGURE 10.6
The precise function
of anting behaviour in
birds is still not agreed
upon. Norfolk, England,
August 2011 (NB).

ejecting their chemicals. Birds using this method spread their wings and tail and do not pick the ants up. Some birds wriggle and squirm, as if dust-bathing, perhaps to incite the ants to swarm over the plumage. In active anting, the insects are picked up in the bill and applied directly onto the plumage. Birds effectively preen with the ants, rubbing them onto the feathers. On occasion, both anting methods are employed at the same time. As they spend much of their time around ants on the ground, Green Woodpecker might indulge in this behaviour more than most other bird species. As they usually start to preen immediately after anting, and in active anting apply the insects directly to the feathers, a plumage maintenance explanation seems to be most appropriate. The chemicals such as formic acid that ants spray when disturbed may act as insecticides, miticides and fungicides that impede plumage parasites.

Roosting

Green Woodpeckers usually roost overnight in tree holes, often previously used nests (see Chapter 13, Cavities, for details of other sites used). They typically roost alone, though occasionally several birds will use different holes in a cluster of trees or adjacent nest boxes. Nest cameras have revealed that birds either squat at the bottom of the chamber or cling to the wall. They will spend the night outside of cavities (Blume 1996; Keicher 2007), probably when there is a lack of suitable sites or because of disturbance at one they frequently use. Recently fledged young may sleep together, tending to squat laterally along branches in cover under the canopies of trees, or they share a hole. Green Woodpeckers may sleep in the same tree as other woodpecker species, although in separate holes (Cramp 1985). A report from Italy of up to ten individuals using wall cavities is exceptional (Brichetti and Fracasso 2020). Rather than any communal instinct, when they roost close to each other, it is probably due to a lack of sites, because of competition for cavities from other species.

These birds typically go to roost just before dark, often calling briefly from nearby beforehand. This may be to enforce territorial claims or to maintain the pair-bond. Despite this, they are generally very circumspect, rarely flying directly to the roost hole but stopping nearby and silently observing before continuing (Keicher 2007). Various factors influence how many times they stop off in intermediate trees. When there is human activity (a forester, farmer, gardener or the like) or a predator in the area, for example, they loiter some distance away before edging closer, from tree to tree. When they do eventually land by the roost, the birds usually look inside before entering, presumably to check for predators or other occupants. It is not unusual for Green Woodpeckers to spend several minutes bobbing their head in and out of the cavity before fully entering. They also often look out again a few times before finally remaining inside for the night.

As already mentioned, light plays a role in the overall activity of Green Woodpeckers, and this is true when it comes to their roost arrival and departure times. They generally leave the site earlier and enter it later on days with clear skies than on overcast days (Keicher 2007). Before departing in the morning, usually soon after sunrise, they typically look out and then withdraw back inside several times. Some birds remain still at the entrance for several minutes; others scan around or flick their tongue in and out. When they eventually leave, they do not loiter by the roost but tend to fly to nearby trees, where

FIGURE 10.7 At dawn a Green Woodpecker peeps out of a nest box where it has roosted overnight. Szár, Hungary, January 2022 (AK).

FIGURE 10.8 Green Woodpeckers are cautious birds. This female has arrived at the nest tree but first hides, checking that all is well, before edging up to the cavity entrance. Nógrád, Hungary, April 2022 (GG).

they may call, before moving away entirely. As when entering the roost, Green Woodpeckers are cautious when leaving. On windy days, however, they can be reluctant to depart – peeping their head out of the cavity entrance, surveying the conditions and situation, before going back inside. Outside the breeding period, birds will visit their regular roosts during the day, particularly in rainy or windy weather, but sometimes even on mild days.

Chapter 11

Movements and Flight

Green Woodpeckers are a non-migratory species; indeed, most individuals are highly sedentary, remaining in their home range throughout the year. In Britain, for example, when breeding and winter distributions were mapped, they were found to be almost identical (Balmer et al. 2013). Once settled in a suitable and successful area, pairs hardly ever relocate. There are known cases of breeding site fidelity for females for 4 years and for males 4.5 years (Glutz von Blotzheim and Bauer 1994). Generally, the longest daily flights made are between roosts – or nests in the breeding season – and foraging areas, but these are often only a few hundred of metres apart.

While they do not make regular seasonal movements, hence not true migrations, good foraging areas, such as gardens, orchards and farmland, may entice Green Woodpeckers in autumn and winter to relocate from the woodlands where they nested. Altitudinal movements, induced by deep snow cover in mountain ranges such as the Alps, are also undertaken (Brichetti and Fracasso 2020). There is evidence that although some populations breeding in uplands descend to lower elevations in winter, a tendency to site fidelity may contribute to mortality in that season (Glue and Südbeck 1997). In lowlands too, individual birds – especially juveniles – may rove farther in hard winters when prey becomes difficult to find. It is notable that Britain's longest recorded distance for a recovered ringed Green Woodpecker (71 km) was during the severe winter of early 1963 (Robinson 2005). Nevertheless, even in the harshest winters Green Woodpeckers do not erupt and move collectively as some other woodpeckers occasionally do. Fledged young leave the natal area in late summer, but sometimes as early as June. These birds must then establish a home range of their own, but typically only move short distances, less than a few kilometres from where they hatched. At such times dispersing juveniles can appear in intermediary habitats, often treeless terrain, where they feed but do not remain to breed in the following spring.

FIGURE 11.1a A male Green Woodpecker swoops down from his nesting tree. Woodpeckers tend to drop rather than rise in flight when leaving trees. Ringmer, East Sussex, England, May 2015 (PW).

FIGURE 11.1b A male Green Woodpecker in flight, his bright yellow rump showing well. Ringmer, East Sussex, England, May 2015 (PW).

FIGURE 11.2 By the end of summer most juvenile Green Woodpeckers have been forced to disperse by their parents, but many do not fly far, initially often just one or two kilometres. Norfolk, England, August 2005 (NB).

Ringing recoveries

Most Green Woodpeckers probably never move far from the place where they fledged. In Britain, the average recovery distance of ringed adults is 1 km and for juveniles 3 km (Balmer et al. 2013), and as mentioned above the longest distance on record is 71 km. In Norway, all ringing recoveries have been within a radius of 20 km from where the birds were ringed (Bakken et al. 2006), and in Sweden, 85% have been within 20 km (SLU Artdatabanken: Artfakta 2020). In Hungary, 96% of recoveries are within 5 km, with two cases involving distances of 40 km and 55 km (Török 2009). In Italy, only two Green Woodpeckers ever have been recovered over 15 km away, at 42 km and 85 km respectively (Spina and Volponi 2008). Yet some recoveries have shown that individuals will move further (Hägvar et al. 1990; Glutz von Blotzheim and Bauer 1994). In the Czech Republic, although recoveries revealed that most birds moved less than 6 km from their breeding sites, one bird was found 107 km away and another 157 km (Cepák et al. 2008). Probably the longest verified distance anywhere was a male in Germany ringed in June and found two years later rearing its young around 170 km to the east of the ringing site (Glutz von Blotzheim and Bauer 1994).

FIGURE 11.3 During post-fledging dispersal juvenile Green Woodpeckers often turn up in ringing camps. This one is about to be released after being caught and ringed in Styria, Austria, in July 2011 (TH).

Sea crossings

Limited dispersal is highly typical of Green Woodpeckers, which is one of the reasons why they are seldom found on islands that are relatively close to the mainland. Despite being quite common in southern England and in France, for instance, they do not breed on any of the Channel Islands between the two (Balmer et al. 2013). In the Netherlands, they are rarely seen on the Wadden Islands despite the archipelago being quite close to the mainland and the high number of birdwatchers there. Yet, in spite of being a chiefly sedentary species that seldom moves far over land, individuals do occasionally fly over stretches of open sea. Though exceptional, some would have had to cross quite long distances, conceivably more than 100 km, to reach some of the islands on which they have been found.

Islands around Britain

Vagrants have been encountered on various Scottish islands, mainly off the west coast, including Bute in the Clyde Islands in 1970, at Kilchoman on Islay in the Inner Hebrides in May 1978 followed by two in 1979, one at Arinagour on Coll in June 1982, and individuals on Mull in May 1980, August 1983, May 1984 and January 1996 (Forrester and Andrews 2007). Individuals from the mainland sometimes visit the larger Welsh islands, but records have become less frequent: there were 15 records up to 1985 on Bardsey, but none since (Pritchard et al. 2021). Green Woodpeckers have been reported on the Isle of Man, but none officially confirmed since the 1950s. The three officially accepted records from Ireland are all from the nineteenth century, but lack detailed documentation: one at Rathmullen, County Donegal in January 1854, an adult male 'obtained' (for which we can read 'shot') at Sallymount, County Kildare on 27 September 1847, and one at Kilshrewly, near Granard, County Longford which has no specific date (Ussher and Warren 1900). There are no confirmed modern records; therefore, Green Woodpecker is listed in Category B (recorded in an apparently natural state at least once up to 31 December 1949) of the Irish list. The absence of any modern-day records has given rise to some scepticism about the validity of the old records, and whether those birds actually crossed the sea unaided, but at present they remain accepted.

Mediterranean islands

Green Woodpeckers have been recorded on islands in the Mediterranean where they are not resident, such as Corsica, Sardinia, Sicily, Malta and Crete.

The sole accepted record from Malta was in May 1903, although two others, with limited details, are in collections: one in the Natural History Museum and one in a private collection (Sultana and Gauci 1982). Green Woodpeckers found in Sicily in recent times are thought to have dispersed from Calabria on the Italian mainland a few kilometres away (La Mantia et al. 2015). The only confirmed record from Crete was a male near Kritsa in April 1978 (Handrinos and Akriotis 1997).

Climate-related dispersal

In Europe, average winter and spring temperatures are known to be increasing and climate change is generally considered to be the main cause. It has been suggested that the gradual range expansion to the north by Green Woodpeckers in Britain may be a consequence of this (Smith 2007; Renwick et al. 2012). Conversely, elsewhere in northern Europe the opposite seems to be the case, with range retractions rather than expansions observed (Wilk 2020). Nonetheless, if such northerly movements of Green Woodpeckers in Britain continue, more occurrences on coastal headlands and on offshore islands may become more frequent (see Chapter 7, Distribution, Population and Trends).

Flight pattern

In flight Green Woodpeckers can appear heavy and bulky. They are strong fliers but not overly agile. Over longer distances, they typically alternate three to four wing beats with brief glides on closed wings, bounding in a so-called typical woodpecker flight. In actual fact, this conception is misleading as many woodpeckers worldwide do not fly in an undulating manner, although Green Woodpeckers do. In his classic *The Natural History of Selborne* Gilbert White wrote that 'wood-peckers fly *volatu undoso*, opening and closing their wings at every stroke, and so are always rising or falling in curves' (White 1906). This flight pattern is not unlike that of Little Owl *Athene noctua*, and in some situations, or on a brief view, the slightly undulating flight of a Golden Oriole could result in that bird being mistaken for a Green Woodpecker (see Chapter 3, Description and Identification).

When flying out from a tree, Green Woodpeckers typically drop low and then level off before rising. They seldom fly above woodland tree-top level, even when crossing open country. Nevertheless, a remarkable observation of one flying up to 100–150 m to visit a hot-air balloon over northern France in

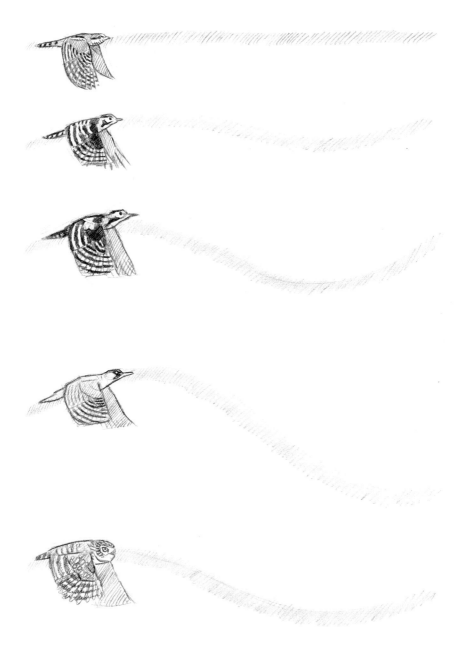

FIGURE 11.4 Comparison of the typical flight patterns of Green Woodpecker and other species (SK). From top to bottom: Eurasian Wryneck, Lesser Spotted Woodpecker, White-backed Woodpecker, Green Woodpecker, Little Owl. Illustrations not to scale.

FIGURE 11.5 A Green Woodpecker takes off. In flight the bright yellow rump and upper-tail coverts are conspicuous. Drnholec, Czech Republic, August 2013 (TG).

June 1993 illustrates just how anything is possible. This high-flying woodpecker approached the basket, called and then left, before returning and alighting on the shoulder of one of the balloonists before flying off again down to the trees below (Fox 1997).

Chapter 12

Breeding

Green Woodpeckers are ready to breed in the spring of their second calendar year. The species is monogamous, with extra-pair mating rare. Pairs attempt one brood per year, and they may start again if a clutch is lost during the laying or incubation stages, but not when a brood is lost. Both parents share the related breeding tasks of incubation, brooding (hence both sexes have an incubation/brood patch), feeding of the nestlings and nest sanitation.

Home range

Green Woodpeckers do exhibit territorial behaviour in the area immediately surrounding their nest and drive off intruders. However, there is no well-defined breeding territory; rather, pairs use a wider home range around the nesting cavity that contains foraging and roosting sites. A study in England calculated the home ranges of a radio-tracked pair to be 151 ha for the male and 15 ha for the female when they were tracked simultaneously, with the male covering a maximum 177 ha (Alder and Marsden 2010). Three radio-tracked birds in Germany revealed the following: the home range of one female was 24.8 ha during the nesting period, but up to 127.9 ha two months later; a male roved over 53 ha, but only used 18 ha for foraging; another male had an autumn and winter home range of 25.8 ha (Ruge 2017). In Norway, where Green Woodpeckers are at the northern edge of their geographical distribution, an average home range of 100 ha has been estimated (Rolstad et al. 2000). Home-range sizes, therefore, are likely to be proportionate to the available resources: most importantly nest and roost sites and food.

Courtship

In mild weather, courtship may commence in late winter with sessions of calling, usually near a potential nesting cavity. One bird will often make repeated flights high over another perched in a tree. Both sexes show cavities to each other and may gently tap or drum by them (Blume 1996). The chosen site is normally located in the softer parts of a living tree. Displays are rather simple, coordinated and mostly take place near the future nest hole. They sometimes appear aggressive and involve excited calling, chases in the air and in trees, head lowering and swaying, and bill touching (Gorman 2004).

Copulation

Green Woodpeckers may mate on a horizontal branch or on the ground. Males offer females food items to entice her, bobbing the head and swaying the upper body. Females may peck at the male's bill in the same way as chicks. The female then lowers her body into a prostrate, submissive posture and makes soft calls. Copulation is brief and involves the male mounting the back of the prostrate female, fluttering his wings and often uttering quiet calls.

FIGURE 12.1 Green Woodpeckers often copulate near the nesting cavity. Emilia-Romagna, Italy, May 2019 (ET).

After mating, both birds usually go to the nest hole or begin to preen. Females also frequently solicit copulation, flying to the male when he is near the nest or squatting by him when he is on the ground (Blume 1996). If an intruding male appears during the pre-nesting period and is repulsed, the paired male usually flies to the female to copulate immediately after the event.

Reverse mounting

Another form of solicitation, which is only occasionally observed, is 'reverse mounting'. This entails a female mounting a male in apparent mock copulation. While it is not uncommon among other birds, this behaviour has not been widely documented in woodpeckers (Gorman 2020b). Although the precise function of reverse mounting is unclear, it is probably a fairly regular part of courtship behaviour, playing a role in strengthening the pair-bond and encouraging further genuine copulations. By mounting the back of the male, a female can probably initiate mating proper (Winkler et al. 1995).

Laying

One egg is laid per day. No nesting material is used: the eggs are simply placed on the bare wood of the chamber floor. Although no material is brought from outside, wood dust and chips produced from excavation can form a cushioning layer, sometimes a few centimetres deep. Film footage from inside nests has shown adults pecking the cavity walls, but whether this is done to deliberately add chips to the floor or simply to widen the cavity for the growing chicks is unclear. Laying commences soon after excavation is finished, sometimes the day following the completion of the nesting cavity (Tracy 1946). Late April into early May is typical, although timing across the range differs accordingly to factors such as latitude and average temperatures in the preceding months. Very cold winters result in delayed springs and consequently later laying dates. A study in Britain found clutches laid as early as March and as late as the end of June (Glue and Boswell 1994).

Clutch size

Complete clutches usually comprise 5–7 eggs, though as few as 3 and as many as 11 are possible (Glutz von Blotzheim and Bauer 1994). These two extremes are, however, uncommon. A review of 49 clutches in Hungary produced the following figures: 1 with 3 eggs, 2 with 4, 15 with 5, 11 with 6, 17 with 7, and 3 with 8, which results in an average of 6 eggs per clutch (Haraszthy 2019).

Eggs

Eggs are typically oval, glossy white and unmarked. They can, however, become soiled from the bodies of incubating adults or stained by debris at the bottom of the dank chamber. They typically weigh around 9 g each of which 7% is shell (Robinson 2005), and measure 30–33 mm in length and 21–25 mm in width. Average sizes calculated from 33 eggs in a collection in the Czech Republic were 31.70 × 23.10 mm (Mlíkovský 2006). Average sizes based upon 94 eggs in Hungarian collections were 30.60 × 22.97 mm, with the largest eggs being 33.00 × 22.20 mm and 31.00 × 24.70 mm, and the smallest 27.00 × 20.40 mm (Haraszthy 2019). Dimensions of eggs within a clutch vary by a few millimetres, with one egg, sometimes two, often visibly smaller. Occasionally one is significantly smaller, even two-thirds the size of the others (Solti 2010). The three images on the opposite page (from the Mátra Museum of the Hungarian Natural History Museum) show clutch sizes and egg dimension variations.

Incubation

Incubation begins when the clutch is either complete or after the female has laid the penultimate egg. It is conducted by both sexes for 14–20 days (Cramp 1985; Glue and Boswell 1994; Glutz von Blotzheim and Bauer 1994) although a period as short as 12 days has been recorded (Tracy 1946). Shifts of 1.5–2.5 hours for each parent are typical. As is the case with most woodpeckers, males tend to incubate overnight. Indeed, male Green Woodpeckers are enthusiastic parents: when an incubating female is reluctant to leave the nest the male can become excitable, changing soft change-over calls to more assertive ones and becoming more animated, bobbing his head and swaying his body.

Hatching and brooding

Green Woodpeckers hatch asynchronously. The chicks that hatch first tend ultimately to be bigger in size than their tardy siblings as they are fed at once. In some broods this can result in one or two underfed nestlings which may not survive. Nestlings are, like all woodpeckers, nidicolous (remaining in the nest for a long time after hatching), naked and at first have their eyes shut. As they lack down feathers they huddle together for warmth and are totally dependent on their parents (thus being termed altricial), who brood them intensively for the first three to five days. The chicks quickly develop bulging, white flanges at the corners of their mouth which help their parents locate them in the dark chamber. The flanges are sensitive and when touched prompt the chicks to

FIGURE 12.2
Seven-egg clutch with
all eggs a similar size.
Cat. No. 03891 (GG).

FIGURE 12.3
Five-egg clutch with
one egg smaller.
Cat. No. 03880 (GG).

FIGURE 12.4
Five-egg clutch with
one egg much smaller.
Cat. No. 03888 (GG).

open their bills. Chicks grow rapidly and their limbs and claws, which are soon needed for climbing the chamber wall, appear oversized compared with their bodies. They are silent for some days but become increasingly noisy as they grow. Most, but not all, nestlings show sexually dimorphic plumage while still in the cavity in the form of red in the malar of males but not the females

(Tracy 1946). They are unable to regulate their own body temperature and so need brooding. As mentioned above, male Green Woodpeckers typically incubate the eggs overnight, and they also tend to stay in the nesting chamber overnight to brood their young. At such times females roost nearby, usually in a cavity. Although Green Woodpeckers incubate their eggs for a comparatively short time, their chicks stay in the nest for a relatively long time. The period from the hatching of the first egg and fledging of the first chick varies greatly: between 18 and 27 days have been documented (Cramp 1985; Glue and Boswell 1994; Blume 1996).

Feeding young

Both parents feed their nestlings with regurgitated food, a pulp of invertebrates, which for the first week or so they thrust into their gullets. Later, most food is given directly from the bill (Blume 1996). When the chicks are

FIGURE 12.5 A female Green Woodpecker feeding a well-grown nestling at the nesting cavity entrance. Emilia-Romagna, Italy, May 2017 (ET).

FIGURE 12.6 As they approach the time to fledge nestlings become rather feisty and parents begin to visit with food less frequently. Emilia-Romagna, Italy, May 2019 (ET).

small, the adults enter the cavity entirely to feed them. At first, young only beg for food when their parents arrive; later they beg almost constantly. After around ten days the nestlings start to clamber up the chamber walls to the entrance hole (Tracy 1946). They stick their heads out, and sometimes their upper body, and are fed there, at the entrance, with the parents remaining outside. The siblings jostle for the best position, often shrieking loudly, and sometimes uttering their first 'laughing' notes, in order to be the first to be fed when food arrives.

They will lunge and stab at their parents' throats in their impatience to be fed, and gradually both parents lose enthusiasm for feeding their demanding and aggressive young, reducing the number of feeding visits. This can be a hazardous time for Green Woodpecker young, as when they lean out of the cavity, making noisy begging calls, they can alert predators to their presence.

Nest sanitation

Although nesting in cavities has many advantages over nesting in the open, it has one major drawback: the enclosed environment is not conducive to hygiene. Both Green Woodpecker adults attempt to keep the nest clean, first by removing eggshells and unhatched eggs and then nestling faeces after feeding. Droppings are enclosed in gelatinous sacs and carried away in the bill, although early in the nesting period (12–14 days) some may be ingested by the parents while in the chamber (Tutt 1956). Later, parents enter the cavity to remove faeces after feeding their young at the entrance. Finally, they stop removing faeces entirely, probably because the chamber is too crowded, or they are unable to enter, and perhaps because the droppings are simply too large. The bottom of chambers invariably become soiled with a mass of wet excrement and food remains by the time the young leave the hole (Tracy 1946).

Fledging

Nestlings require considerable attention from their parents in order to grow and finally fledge. As young become more demanding, even pecking at their parents' heads and throats when they arrive, both parents start to withhold food. They then reduce their visits to the nest and finally stop feeding at the nest entrance entirely. Rather than landing at the nest, parents perch nearby, sometimes in another tree, and make low, soft calls, aimed at enticing their young to leave the cavity. These actions motivate their offspring to leave the cavity. It is also possible that conditions inside the nest, where chicks are cramped together, among the waste which has accumulated towards the end of the nestling period, also contributes to the decision to fledge (Tutt 1956). Occasionally, young that are ready to fledge and which are not being regularly fed by their parents remain in the cavity for up to a week before they finally leave. This usually happens when there are strong winds or heavy rain.

Once out of the cavity the brood splits between parents, who feed them for up to seven weeks (Glutz von Blotzheim and Bauer 1994). There is no obvious structure to this split, parent birds remaining with any number of young. Neither is the division of offspring done according to gender; a male parent may stay with male or female young, or a combination of the sexes, as do adult females. The two groups may meet, as their food resources are often in the same area, but they do not readily interact.

They then begin to disperse short distances. Some fledglings will stay together for weeks, even roosting together, squatting on branches in tree canopies until they find suitable separate roosting cavities. This is one of the

FIGURE 12.7 A Green Woodpecker nestling tries out its long tongue a few days before finally fledging. Emilia-Romagna, Italy, May 2019 (ET).

FIGURE 12.8
By the end of summer most juveniles are independent of their parents, which no longer feed nor tolerate them, and they begin to forage alone. Norfolk, England, August 2005 (NB).

rare periods when Green Woodpeckers can be seen in numbers. Fledglings on the ground will occasionally flick their tongue in and out of the bill, lashing it around in the air and across the ground. Perhaps they are practising how to use this remarkable tool?

Breeding success

The success of a species is a difficult concept to evaluate (Fuller 1982), but as is typical for the majority of cavity-nesting birds of all families, Green Woodpeckers generally have high breeding success rates. Up to seven, but usually two to five, young can be reared. A study in Britain of 252 nests found that for the combined laying and incubation periods the relative success of nests (based upon at least one egg hatching) was 86.8% (Glue and Boswell 1994). In the same study the success of pairs in rearing at least one young bird through to fledging was estimated to be a very impressive 95.3%.

FIGURE 12.9 Change-over at the nest. A male waits ready to enter the cavity as the female exits. Nógrád, Hungary, April 2022 (GG).

Breeding failure

Reasons for nesting attempts failing include sudden weather events, predation of one (or both) of the pair or of the clutch or brood, and displacement by cavity-competitors such as Common Starling *Sturnus vulgaris*. In Britain, Grey Squirrels *Sciurus carolinensis* also compete for cavities (Newson et al. 2010). As Green Woodpeckers often nest in places frequented by people (gardens, parks) disturbance (intentional and unintentional) is another, not uncommon, cause of breeding failure. Failures for Green Woodpeckers, as in other woodpecker species, tend to be slightly higher during the laying and incubation periods than during the nestling period (Glue and Boswell 1994).

Chapter 13
Cavities

The ability of woodpeckers to excavate cavities in trees sets them apart from almost all other birds. A typical woodpecker cavity consists of an entrance hole of about the same diameter as the woodpecker's body and a short entry passageway which then leads down into a wider chamber.

Even though Green Woodpeckers are primarily terrestrial foragers, they, too, are highly adept at creating holes in trees. Cavities provide more shelter from wind and rain than open nests do and also better protection from predators, as a result of which cavity-nesting birds tend to have higher levels of

FIGURE 13.1
A snug fit. As is typical for woodpeckers, the diameter of the nest entrance hole is roughly the same size as the bird's body. Nógrád, Hungary, April 2022 (GG).

FIGURE 13.2
An adult male Green
Woodpecker looks
out from his nesting
cavity entrance.
Emilia-Romagna, Italy,
May 2019 (ET).

breeding success than birds that nest in the open. Green Woodpeckers are no exception (see Chapter 12, Breeding). The species is, like all true woodpeckers, a primary cavity excavator – in other words, it makes its own holes. Natural cavities in trees, meaning those not made by the woodpeckers themselves, such as hollows, are rarely used. Those birds and other wildlife which nest, roost or den in tree cavities but do not themselves excavate them are known as secondary cavity users (see below).

Nesting cavities

Both Green Woodpecker sexes excavate. Most cavities for use as nesting sites are made in spring, generally in March and April, the precise timing being influenced by location (latitude and elevation) and weather. Some exploratory excavations may be done in the previous autumn. Several cavities are started: some finished, some apparently not completed. One will be used for breeding, others for roosting, others not used at all. The pair work together to complete the site finally decided upon, with males tending to do most of the final excavation (Glutz von Blotzheim and Bauer 1994). A successful nesting cavity from which young fledge without interference may be used again, occasionally

FIGURE 13.3
Newly excavated cavities have clean entrance rims. Note that another hole was started above the completed one, but not finished: probably owing to the wood there proving to be too hard. Nógrád, Hungary, April 2022 (GG).

for several years, though not necessarily consecutively. In areas where there is a lack of suitable trees for cavities, reusing old ones may be more common (Henderson and Henderson 2002). However, all 33 cavities documented in a study in Hungary by the author were newly excavated, none having been created in previous years and being reused (Gorman 2020c). New cavities can be recognised by their entrances having clean edges with no renewed tree growth, and light-coloured wood (Gorman 1995).

Excavation

How pairs decide where to make a cavity is unclear. Green Woodpecker pairs are generally site faithful, so new holes are mostly located close to old ones, sometimes in the same tree. Green Woodpeckers make cavities in living, dying and dead trees, but it is often stated that they are less reliant upon dead trees than are most woodpecker species (Hägvar et al. 1990; Smith 2007). For example, a review of 146 nests in France found 28.1% were located in 'unhealthy' trees and 71.9% in 'healthy' trees (Grangé and Fourcade 2019). Yet exactly when a tree is alive, healthy, unhealthy or dead, and degree of decay, are all difficult terms to define. Trees with internal decay, which is often

FIGURE 13.4 Making holes in living trees can entail a risk. This adult male excavated a nesting cavity in a living White Mulberry *Morus alba* and became soaked by the sap which oozed from the wounded trunk. Note also that the usually dark facial-mask is rather pale, which suggested that this individual might be a *viridis* × *sharpei* cross, especially as the border of the known hybrid zone is less than 250 km to the west. It was decided, however, that the bird was not a hybrid but simply possessing aberrant plumage. La Garde, southeast France, June 2019 (JMB).

not obvious to humans as there may be no external indications of rot, are easier to excavate, and most woodpeckers chose such trees because of this. Trees that are solid externally but relatively soft internally, usually owing to fungal activity, are typically selected. Ultimately, for most woodpeckers, probably including the Green Woodpecker, a trade-off seems to exist between location and how easy the wood is to excavate.

Green Woodpeckers do not have an affinity with any particular tree species (Riemer et al. 2010). A combination of factors, such as location, condition and relative size of each tree, rather than its species, results in it being selected for the creation of a cavity.

During the first stage of excavation, wood chips and debris are tossed directly out from the cavity entrance with sweeping and shaking actions of the head. They are not carried away but accumulate below the tree (Gorman 2015). This removal of debris is typically repeated at intervals of between one and five minutes.

When excavating a nesting cavity, Green Woodpeckers are rather cautious – much more so, for example, than Black Woodpeckers, which will often continue to work on their holes while being watched. If an excavating Green Woodpecker is aware that it is being observed (by a researcher, photographer or just a curious observer) it may put its head out of the hole, or leave the hole and perch nearby, either moving its head and gently calling, apparently listening, or simply freezing. When flying from the nest during excavation, Green Woodpeckers may either leave the immediate vicinity or land on a nearby tree, or a branch on the nest tree, and freeze there, sometimes remaining still for ten minutes or so.

Duration of excavation

The excavation of nesting cavities can take anything between 15 and 30 days to complete (Labitte 1953; Glutz von Blotzheim and Bauer 1994). The actual time taken is difficult to assess, as work is not always continuous. Some cavities are started and then abandoned, others returned to and completed in the same season, and yet others revisited a year or more later. Completed cavities that are to be used are guarded before laying commences, both adults taking turns to sit by the entrance or squat in the chamber to deter any potential usurpers.

Cavity location

Green Woodpecker nesting cavities are usually on the main trunk of a large tree, in a foliage-free section below the canopy and with a clear flight path to the entrance. Branches are only occasionally used. For example, all 33 nest holes documented by the author in a study in Hungary were in trunks (Gorman 2020c). A review of 197 nests in France found 86.8% were in trunks and just 13.2% in branches (Grangé and Fourcade 2019). With woodpeckers in general, the mean diameter of a nesting trunk or bough decreases with the size of the woodpecker. In Britain, Glue and Boswell (1994) found a mean diameter of 39.4 cm (range 31–46 cm; n = 4) for trunks and branches housing Green Woodpecker nests. In Hungary, the mean diameter of nesting tree trunks documented was 43.1 cm (range 36–55 cm; n = 33); the most frequent trunk diameter was 45 cm (n = 6: 18%) (Gorman 2020c).

FIGURE 13.5 This nest hole in a poplar tree in lightly wooded country was placed in a typical spot, on the main trunk with an unobstructed flight path towards it. Fejér County, Hungary, April 2022 (GG).

Hole height

Entrance-hole height varies considerably, but between 2 m and 6 m is typical (Glue 1993; Glutz von Blotzheim and Bauer 1994; Solti 2010). A study of 33 nest holes in Hungary by the author found heights ranging from 2 m to 9 m (Gorman 2020c). The most frequently documented height was 5 m (n = 7: 21.21%), with the mean height of 5.6 m. Exceptionally, holes can be as low as just 0.36 m above ground level and as high as 24 m (Henderson and Henderson 2002). Using nest-card records in Britain, Glue and Boswell (1994) calculated a mean height of 4.0 m (range 0.6–15.2 m). The height of the nest hole as a percentage of the total height of the tree, where it was known, was 46.6% (range 12.5–93.3%; n = 15). Sometimes a new cavity is excavated in a tree where one already exists, and in such cases the new entrance is generally placed below the old one.

FIGURE 13.6 Once their cavity is completed Green Woodpeckers guard it from usurpers. Such holes are highly sought after by many other species which are unable to make them themselves, so-called secondary cavity users. Emilia-Romagna, Italy, May 2019 (ET).

FIGURE 13.7 Entrances of reused cavities in living trees tend to become irregularly shaped and look 'natural' as the wood heals and regrows around the rim. Toulon, France, May 2020 (JMB).

Entrance shape and size

The shape of entrance holes varies, but most are circular and between 5 and 7 cm in diameter (Cramp 1985; Glutz von Blotzheim and Bauer 1994). A study in Britain calculated a mean diameter of 5.3 cm (Glue and Boswell 1994). On occasion entrances are slightly oval-shaped, being up to 7.5 cm vertically, but seldom as elliptical as those of Black Woodpeckers. In a study of 33 holes in Hungary, 28 were circular in shape, four vertically oval and one horizontally oval (Gorman 2020c). The entrances of old cavities that are reused can be irregularly shaped owing to the tree regrowing (healing) around the rim.

Nest chamber

The internal dimensions of nesting chambers vary. Widths of between 15 and 20 cm are typical (Glutz von Blotzheim and Bauer 1994), but depths can differ significantly. A study in Britain found chamber depth ranged from 20 to 107 cm, with a mean figure of 38.1 cm (Glue and Boswell 1994). Depths around or above 100 cm are certainly not usual.

Orientation

Several studies in different places around the world have demonstrated that woodpeckers' cavity-entrance orientation is typically non-random (Landler et al. 2014). This suggests that woodpeckers choose certain compass directions for the entrance holes because these enhance breeding success. Woodpeckers inhabiting higher latitudes tend to prefer southerly directions, which implies that climate and temperature are factors which drive their choice. A suitable temperature inside the chamber is essential, and entrances facing southwards will typically receive more solar heating and be less exposed to cold winds.

FIGURE 13.8 Nest hole in an old apple tree in a rural garden. The entrance was 2.5 m above the ground and faced south-west. Nógrád, Hungary, April 2022 (GG).

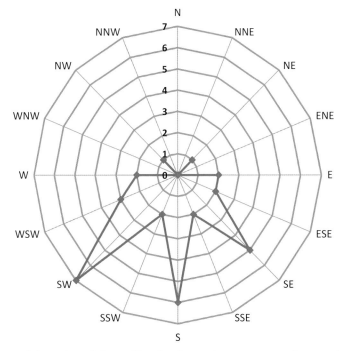

FIGURE 13.9 Orientation of Green Woodpecker cavity entrances as frequencies of cardinal points based on thirty-three nests in Hungary. It can be seen that orientation towards the south dominated.

Yet other factors, such as location and ease of excavation, are probably also involved: the softer parts of a tree may not be on its south-facing side. When cavity trees are on hillsides, entrances generally face down the slope, although this may not be in a southerly direction. It has been suggested that the reason for this is based on the need for a 'good outlook' (Stenberg 1996). Then again, cavities in trees in open landscapes are more likely to be situated facing northwards than are those inside woodlands. It has been proposed that this is in order to lessen overheating by the sun (Stenberg 1996).

Data on the orientation of Green Woodpecker cavity entrances are limited, but a study in Hungary by the author found them to be non-random (Gorman 2020c). Of 33 cavities documented, 22 (66.67%) faced southwards (SE, SSE, S, SSW, SW). The most frequent alignment was SW, with seven (21.21%) cavities. Four cavities were orientated towards the east (E, ESE), five westwards (WSW, W, WNW), one cavity was orientated towards the NW and one to the NE. Hence, the cavity orientation was non-random and significantly biased towards the south, with a mean direction of 187 degrees clockwise from north. In addition, a similar study in Hungary on its close relative, the Grey-headed

Woodpecker, also found that the direction their holes faced was non-random, and that a southerly orientation prevailed (Gorman 2019). A review of 111 Green Woodpecker nests in France found 15.3% orientated to the north, 18% to the east, 22.5% to the west and 44% to the south (Grangé and Fourcade 2019).

Roosting cavities

Green Woodpeckers roost mostly in cavities in trees which they previously used as nesting sites, although natural tree hollows that have been adapted and 'improved', and unoccupied Black Woodpecker holes are also sometimes used (Blume 1996). When they inspect Black Woodpecker cavities, it is probably to consider roosting in them. Away from trees, the burrows of European Bee-eaters *Merops apiaster* in sand-pit walls are also occupied after having been enlarged. Nest boxes and sometimes holes in utility-poles and buildings are also exploited for this purpose (Cramp 1985; Brichetti and Fracasso 2020). In June 2002 in Slovenia a bird was observed making and using a roosting hole in the wooden shutters of a castle, behaviour described as rare (Tomazic 2002).

Occasionally, cavities excavated specifically for roosting are made outside the breeding season, in autumn and winter (Gorman 2011; Grangé et al. 2020). The woodpeckers may do this simply in response to a lack of cavities in which to roost. It is also possible that the individuals that make such cavities are recently independent young in their first winter which are preparing for their first breeding season in the forthcoming spring.

FIGURE 13.10
A Green Woodpecker entering its night-time roost in the attic of a house. Balf, Hungary, January 2022 (BS).

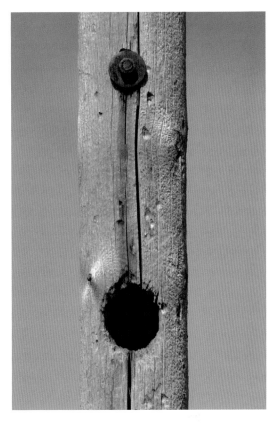

FIGURE 13.11
Roosting cavity excavated and used by a Green Woodpecker in a wooden utility-pole. The round shape of the entrance is typical for this species. The diameter of this hole was approximately 7 cm. Heves County, Hungary (GG).

A noteworthy record from Hungary concerned an individual that regularly roosted in a hole at the top of a concrete pillar (Rozgonyi 1999). All in all, a general lack of tree cavities, together with the intense competition for those that exist (from other woodpeckers and other wildlife), means that unusual sites not placed in trees on occasion have to be used as roosts.

Other cavity users

When they make cavities, woodpeckers create new niches, including in places where some naturally formed cavities are present. As primary cavity excavators they provide (albeit inadvertently) opportunities for many secondary cavity users. A wide range of birds, mammals and insects benefit from the holes that Green Woodpeckers make.

Birds

In grassland and farmland landscapes across southern Europe, where large trees with natural hollows are often scarce (particularly as they are often removed by farmers and landowners), Green Woodpecker holes are often used by European Rollers *Coracias garrulus* (Bouvier et al. 2014). A lengthy list of other birds that do likewise includes Hoopoe *Upupa epops*, Common Starling *Sturnus vulgaris*, Great Tit *Parus major*, Eurasian Nuthatch *Sitta europaea*, Tree Sparrow *Passer montanus*, House Sparrow *Passer domesticus* and the non-excavating picid the Wryneck *Jynx torquilla* (Bijlsma 2014).

In Britain, the non-native Rose-ringed Parakeet *Psittacula krameria* also uses Green Woodpecker cavities for nesting (Butler et al. 2013). Just how important woodpecker cavities can be for other birds can be illustrated by the following two records. First, an observation from May 1949 in Hungary when five presumed Green Woodpecker holes, located between 2 m and 4 m above ground in a single White Poplar *Populus alba*, hosted two breeding pairs of Tree Sparrows, a pair of Eurasian Nuthatches and a pair of European Rollers, as well as a pair of Green Woodpeckers (Janisch 1954). Second, in the summer of 2021 in the city of Olomouc, Czech Republic, the first two urban breeding records of Scops Owl *Otus scops* in the country were documented. One pair used a cavity excavated in the wall of a hotel and the other a cavity in a school building, both created by Green Woodpeckers (Grim et al. 2022; Kovařík et al. 2022). These records not only illustrate how holes made by Green Woodpeckers can benefit other species, but also how their movement into settlements can also facilitate the same for others.

FIGURE 13.12
Hoopoe nesting in an old Green Woodpecker hole. Bologna, Italy, May 2021 (ET).

FIGURE 13.13 A Scops Owl brings food for its chick, the nest inside a cavity made by a Green Woodpecker in the wall insulation of a hotel. Olomouc, Czech Republic, July 2021 (TG).

Mammals

Some woodland bats rely on woodpeckers as providers of cavities for their roosts. In Britain, for instance, Green Woodpecker holes are often used as roosting sites by the threatened Bechstein's Bat *Myotis bechsteinii*, a woodland species which is vulnerable across Europe (Alder and Marsden 2010). The exacting conditions that most bats require for their maternal nurseries probably mean that few woodpecker cavities (of any species) are ideal for that purpose. Other mammals that make use of Green Woodpeckers holes, either as breeding, denning or sleeping sites, include Red Squirrel *Sciurus vulgaris*, Garden Dormouse *Eliomys quercinus* and Edible Dormouse *Glis glis*. No doubt many other small and medium-sized animals occasionally also do so across this woodpecker's range, but information on the subject is somewhat scarce. Insects such as wasps, hornets and bees likewise utilise woodpecker cavities. However, not all Green Woodpecker cavities will be suitable for reuse – by the woodpeckers themselves or other wildlife. They may deteriorate, be too soiled by containing excessive excreta or be known to predators.

FIGURES 13.14a, 13.14b, 13.14c
This male Green Woodpecker discovered a pair of Common Starlings using its cavity and promptly proceeded to remove the nesting material of the intruders. Bologna, Italy, May 2010 (FB).

Chapter 14

Tracks and Signs

Like most woodland wildlife, Green Woodpeckers leave signs that reveal their presence and offer clues to their activities. In periods when they are less vocal, these signs can indicate whether the birds inhabit an area or not. Obviously, nest holes in trees are one of the most noticeable indicators that woodpeckers are in an area. Hole-entrance size is significant, but can be difficult to judge from the ground, and entrance shape is also a good indication. Green Woodpeckers tend to make round entrance holes, but not always, as they can also be slightly elliptical. For more on this see Chapter 13, Cavities.

FIGURES 14.1a and 14.1b Green Woodpecker nesting cavity entrances (left) are usually round, whilst those of Black Woodpeckers (right) are typically oval and often teardrop shaped. Although smaller than those of their larger cousin, size can be hard to judge with the naked eye when holes are high-up in trees. The Green Woodpecker hole here was approximately 6 cm in diameter, the Black Woodpecker hole approximately 10 cm wide and 12 cm high. Both images taken in the Vértes Hills, Hungary (GG).

Green Woodpeckers mainly forage on the ground, in grassy areas such as woodland glades, clearings, pastures and meadows. Many tracks and signs can at first appear chaotic, but if one inspects such places carefully a picture often develops of activity that facilitates the identification of the animal responsible (Wenger 2021a). Still, not all the signs that Green Woodpeckers leave will be assignable to the species as, for example, Grey-headed and Black Woodpeckers will also feed on the ground, as well as birds from other families. Therefore, as much time in the field as possible, with at all times an attitude of circumspection, is needed to gain experience and acquire knowledge of this often-challenging subject.

Wood chips

Like all woodpeckers, Green Woodpeckers produce woody waste when making cavities in trees. Wood chips from excavation accumulate at the base of tree trunks, as they are not carried away but simply tossed out from the entrance hole. Chips and dust alone are seldom sufficient evidence to identify which species was responsible for the cavity, although the dimensions of wood chips can offer clues to the species involved (Gorman 2015). Those made by Green Woodpeckers can be up to 5 cm long; wood chips bigger than that have usually been produced by Black Woodpeckers (Gorman 2010). In addition, chip size is dependent on the structure of the wood: that is, whether it is long-grained or short-grained and whether the fibres are tough or not. When Green Woodpeckers reuse a cavity, few, if any, wood chips lie on the ground below it.

Foraging holes

Although Green Woodpeckers mainly look for food on the ground, they do forage in trees, often in areas of deadwood, and even occasionally on cliffs and buildings (see Chapter 15, Foraging and Food). Green Woodpeckers will 'beat-up' rotten tree stumps to get at insect prey – a habit they share with Black Woodpeckers – but they work more precisely and the debris they leave is less messy, and the discarded wood chips are smaller than those made by their larger cousin. Green Woodpeckers only infrequently create large, deep cavities in trees when looking for food; rather, they strip, prise and chisel at already loose bark. Generally, the holes and marks they leave on and in timber when searching for invertebrates are not diagnostic, being practically indistinguishable from those made by the other woodpeckers with which they are sympatric.

The holes that Green Woodpeckers make in terrestrial substrates when searching for ants, however, are another matter. These can often be attributed to the species, although in southern France where the Iberian Green Woodpecker coincides with it, the foraging work of the two close relatives is probably inseparable. In some situations, holes made in the ground by Grey-headed Woodpeckers, and even Black Woodpeckers, cannot be ruled out either (Gorman 1995).

Pits in the ground

Small, narrow, conical foraging pits of 2–3 cm in diameter and 5–15 cm deep are dug in grassy areas to access the small ants that live in colonies in the soil below (Gorman 2015). In winter, when ants move further underground, the pits are deeper, with entrances up to 11 cm in diameter; those of Grey-headed Woodpeckers are smaller at around 7.5 cm (Löhrl 1977). Green Woodpeckers tend to work efficiently and neatly, so pit entrances are roughly circular with well-defined rims. Soil or other debris is not tossed far, so very little is scattered in the surrounding area. Rather, prised up clods of soil, sods of turf or clumps of moss are more typical. This orderly foraging style can be contrasted with that of Black Woodpeckers, which often toss and scatter debris over several metres. Additionally, various crows, especially Rooks *Corvus frugilegus* and Common Starlings will peck and prod into the ground in search of food. Whereas Green Woodpeckers almost always search for food alone, these birds do so in flocks

FIGURE 14.2 Green Woodpeckers will dig small round pits in the ground with their bill, through grass and soil, to access ants and other hidden prey. Czech Republic, August 2017 (TG).

and thus leave many signs over a large area, with much discarded debris. Foraging pits might also be confused with those made by the Eurasian Badger *Meles meles* and some hedgehog species (Erinaceinae). Hedgehogs in particular make quite precise holes in the ground when searching for earthworms, and these can look very much like those made by Green Woodpeckers (Wenger 2021a). Other signs left by these mammals (pawprints, scat), or the presence of woodpecker droppings, can help in establishing who dug them (Gorman 1995; 2015).

Holes in snow

In winter Green Woodpeckers will clear away snow to access prey in the soil beneath. The cleared patches are typically round with a fairly well-defined rim and are 5–15 cm in diameter. Conical pits, like those mentioned above, are then dug into the ground in the cleared area.

Holes in ant mounds

Conical, funnel-shaped holes in the mounds (ant hills) of wood ants (*Formica* species) are a sign of Green Woodpecker foraging. These are typically 5–10 cm (but up to 20 cm) in diameter and 10–60 cm deep (Gorman 2015). Some are more like tunnels than funnels as they can reach depths of 80 cm (Wenger 2021a), and one an impressive one metre long has also been measured (de Bruyn et al. 1972). There are usually several smaller holes on the same mound,

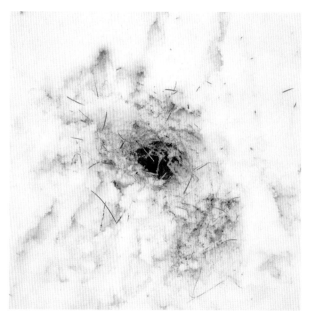

FIGURE 14.3
Green Woodpeckers will dig with their bill through deep or hard snow to access ants and other prey hidden in the ground beneath. This pit was 8 cm in diameter. Budapest, Hungary, December 2012 (GG).

FIGURE 14.4a Ants swarm on the top of their mound and attempt to repair it after it has been opened up by a Green Woodpecker. Vértes Hills, Hungary, April 2022 (GG).

FIGURE 14.4b Foraging holes in a long-established wood ant mound. As can be seen here, Green Woodpeckers will repeatedly visit such sites. Žarnovica, Slovakia (TJ).

FIGURE 14.5 Foraging holes, approx. 20 cm diameter, made in a frozen ant mound (perhaps *Formica pratensis*). Vértes Hills, Hungary, December 2021 (GG).

too, where a bird has probed with its bill to find the most productive spots. As with pits made in the ground, Green Woodpeckers work tidily, tending to focus on precise spots and creating very little scattered debris. Such holes are visited repeatedly when they contain good numbers of prey, and they are often easier to find and recognise than the holes Green Woodpeckers make in the ground.

Grey-headed Woodpeckers also visit ant colonies and the signs they leave are similar to those of their relative. Black Woodpeckers, on the other hand, tend to be more untidy when tunnelling into mounds. Their habit of swishing the bill scatters debris laterally in all directions, and the pits made are seldom as precise as those of Green Woodpeckers (Gorman 1995, 2015). Another bird that delves into ant colonies is the Hazel Grouse *Bonasa bonasia*. The best way to separate its foraging work from that of Green Woodpeckers is to look for its droppings, which are smaller and typical in form for a gamebird.

Some mammals also raid ant mounds and the holes they make in them can be confused with those of Green Woodpeckers. However, when the likes of Badgers, Brown Bears *Ursus arctos* and Wild Boars *Sus scrofa* do so, the site is invariably more trashed, with bigger and deeper holes made and more debris evident. As with pits in the ground, the presence of other evidence at the scene, such as tracks and droppings, can help in identifying which animals foraged there. Although such holes can be found all year-round, Green Woodpeckers (and indeed other woodpeckers) tend to visit ant mounds most frequently in winter.

Droppings

The droppings of most woodpeckers are hard to find and identify to species. As Green Woodpeckers often feed in open areas, such as on garden lawns and sports fields, their droppings are generally easier to find than those of their relatives that inhabit the depths of forests. They are also quite characteristic in shape, colour and content. They are cylindrical and elongated, usually 10–20 mm in length and 4–5 mm in diameter. When fresh they are enclosed in a greyish-whitish membrane and resemble soggy cigarette butts. As they dry, they become whiter owing to a coating of uric acid. Once dried out, they become brittle and delicate

FIGURE 14.6 A frozen Green Woodpecker dropping, approx. 2 cm long, on an ant mound. Vértes Hills, Hungary, December 2021 (GG).

and easily crumble if picked up. They are usually composed of brownish-reddish, ash-like fragments of undigested ant exoskeletons. At regular feeding sites they can be found in small piles, which are not latrines but simply accumulations. Sometimes just one or two droppings are present. Green Woodpeckers also regurgitate pellets that, like their droppings, are mostly composed of the hard, shiny indigestible exoskeletons of ants. Not all ant exoskeletons are the same, however. Those of *Myrmica* and *Lasius* ants apparently differ significantly in their crushability. *Myrmica* fragments are tougher and harder and so are ejected in pellets, while those of *Lasius* species are seemingly easier to digest and consequently end up in the faeces (King et al. 1973).

Footprints and foot arrangement

Green Woodpecker footprints are hard to find as these birds do not readily step into mud and those left on dusty tracks quickly disappear. Prints in ant mounds are typically ill-defined, as the fragile pine-needle surface invariably quickly collapses around them. Therefore, the best time to look for Green Woodpecker footprints, and also traces of the wings and tail, is usually after a fall of snow. Nevertheless, the clarity of any prints will depend upon the ground conditions: the best substrate is hard, crisp snow rather than soft or melting, as it maintains the shape and dimensions of the feet better. In the rare cases that Green Woodpeckers' 'walking' trails are found, two almost-parallel lines indicate a hopping gait (Wenger 2021a).

Green Woodpeckers have four toes on each foot, as do nine of the ten other woodpecker species that occur within their range. The exception is the Three-toed Woodpecker (see Chapter 2, Anatomy and Morphology). The four toes are usually positioned in a zygodactylous arrangement: digits 2 and 3 pointing forwards in parallel and digits 1 and 4 backwards. This layout leaves a K- or X-shaped footprint. The toes leave slender imprints, with the front two toes closer together than the hind ones (Wenger 2021b). The inner two toes (digits 1 and 2) are shorter and slimmer than the outer ones (digits 3 and 4), indeed the small hallux (digit 1) often does not leave an impression. The claws are long and curved, but they do not often register well, either. On the other hand, the metatarsal pad – the area in the centre of the foot where all the toes meet – always touches the ground and generally leaves the largest impression of the print. An excellent study on this subject by Wenger (2021b) produced an average Green Woodpecker footprint size as follows: width 1.7 cm, width without claw marks 1.2 cm, track length 6.3 cm, and track length without claw marks 4.5 cm.

FIGURE 14.7 Green Woodpecker tracks in snow. Budapest, Hungary, December 2012 (GG).

Yet the irregularity of tracks often means it is difficult to ascertain the precise species that made them and sometimes even to be sure that the foot was zygodactyl. In any group of tracks found, some will be indistinct and impossible to interpret, but one of two are usually clear enough to do so with some accuracy. It should also be remembered that within the range of the Green Woodpecker some other birds, such as cuckoos and owls, have similar zygodactylous feet, but typical owl feet, for instance, leave thicker impressions than those of woodpeckers due to their plumper toe-pads. Yet, as these other birds do not visit the ground as often as Green Woodpeckers do, and even less so when there is snow, in most footprint scenarios they can effectively be ruled out as confusion species.

Use of anvils?

There are anecdotal reports of Green Woodpeckers using 'anvils': small holes or crevices in trees where hard nuts or conifer cones are wedged so that they can be more easily opened and the contents extracted. If this feeding behaviour is indeed performed, it is certainly rare and it is doubtful that any signs left could be distinguished from those made by Great Spotted and Syrian Woodpeckers, two species that frequently create and use such 'woodpecker workshops'.

Chapter 15

Foraging and Food

Differences in the foraging habits of woodpeckers are reflected in their physiques (see Chapter 2, Anatomy and Morphology). The 11 species that occur in Europe can be broadly placed in three groups in terms of their diet: omnivores, forest invertebrate-eaters and ant-eaters (Mikusiński and Angelstam 1997). The Green Woodpecker is one of the five species, along with Black, Grey-headed and Iberian Green Woodpeckers and the Wryneck, that feeds predominately on ants. It is above all a terrestrial forager with a narrow diet. Indeed, its limited range of food consists mainly of ground-, soil- and mound-living ants. Though specialised in feeding on ants Green Woodpeckers, like most other picids, can be opportunistic and will exploit other food resources. Soil-dwelling invertebrates such as earthworms are taken as well as the larvae of bark- and wood-living beetles, woodlice, flies, carpenter moth larvae and pupae, spiders, caterpillars, snails, and wasp and bee grubs (Cramp 1985). Nevertheless, the species takes fewer arthropods from trunks and branches than most woodpeckers in its range and in this regard is like most of its relatives in the *Picus* genus (Winkler et al. 1995).

Ants

When ants (Formicidae) are abundant, Green Woodpeckers eat little else (Korodi Gal 1975; Pechacek and Krištin 1993; Raqué and Ruge 1999). The analysis of their stomach contents has revealed that millions of ants can be consumed at productive sites (Cramp 1985). These invertebrates are eaten in all stages: eggs, larvae, pupae and adults, often mostly workers but also queen ants (King et al. 1973). They are collected directly from the ground or dug from their colonies in soil and mounds. Despite this, the effect of Green Woodpecker predation on ant populations is thought to be limited

FIGURE 15.1 A Green Woodpecker on a garden lawn with insect prey. This individual is moulting from juvenile into adult plumage. Norfolk, England, September 2008 (NB).

(de Bruyn et al. 1972). These woodpeckers apparently do not take all the ant species within their home range, which is probably related to habitat structure affecting accessibility (Alder and Marsden 2010). Small, mound-living *Formica* (wood ants), mainly turf-dwelling *Myrmica* (red ants) and *Lasius* (moisture ants) species, which typically inhabit open short-grassed areas, usually form the staple diet. Which ant species (and other invertebrates) are sought also depends upon variable factors such as the season, prevailing weather, temperature and their local abundance.

In Britain, as in much of Europe, the Yellow Meadow Ant *Lasius flavus* is common and widespread and thus a favourite prey species, especially during the breeding season (Alder and Marsden 2010). In the southern Netherlands, a study of the contents of both adult and nestling Green Woodpecker droppings found that the diet of birds living in open areas consisted mainly of Black Garden Ants *Lasius niger*, in both summer and winter, while that of those living in more wooded habitats was more diversified but consisted mainly of *Lasius platythorax*. In both landscapes the proportion of *Lasius* was smaller in winter than during the breeding season. Mound-building *Formica* ant species were found in the winter diet of the woodpeckers using woodlands but constituted

FIGURE 15.2 This 'daredevil' ant on the top of the woodpecker's head is safe for now. Czech Republic, June 2018 (TG).

only 14% of the total number of food remains identified (Klosters et al. 2014). Across the Green Woodpecker's range, other ants known to be frequently predated include *Formica polyctena, F. nigricans, F. cordieri, F. fusca, F. exsecta, F. pressilabris, F. picea, Lasius alienus, Manica rubida, Myrmica rubra, M. schencki, M. scabrinodis, M. laevinodis, Tetramorium caespitum, Serviformica rufibarbis, Camponotus herculeanus, C. pubescens* and *C. vagus* (King et al. 1973; Glutz von Blotzheim and Bauer 1994; Blume 1996; Raqué and Ruge 1999).

Seasonal differences

The availability of ants, especially in winter, considerably influences the distribution, population and breeding success of Green Woodpeckers. In very hard winters, the upper layer of soil and the outside surface of ant mounds can freeze, hampering the woodpeckers' ability to dig for prey. At such times Green Woodpeckers may move from exposed open areas to warmer woodland interiors to forage for ants that have their colonies in rotten tree trunks and stumps. The mounds of wood ants inside forests are also more frequently visited, probably because being under the cover of trees they are afforded some shelter from heavy snow. As wood ant mounds can be large, they are also more accessible than the smaller ones of ant species that live in exposed habitats such as meadows (Rolstad et al. 2000). In periods of harsh weather such colonies can be a crucially reliable source of food. One study found that the Red Wood Ant *Formica rufa* can form the bulk of the Green Woodpecker's diet in winter, while *Lasius* species such as Black Garden Ant and Jet-black Ant *Lasius fuliginosus* are eaten in spring and summer (de Bruyn et al. 1972). In continental Europe (but not in Britain as they do not occur there) various species of large carpenter ants (*Camponotus*) may also be sought in winter.

Plant food

Occasionally non-invertebrate prey is eaten by Green Woodpeckers. Berries, seeds and windfall fruit such as apples, pears, cherries and grapes are sometimes consumed (Glutz von Blotzheim and Bauer 1994; Gorman 2004). Locally, such vegetable matter may become a frequent and probably important part of the diet, particularly in winter. An account from a garden in England detailed how the woodpecker ate apples on the ground, attacking them vigorously at one end and excavating them around the core, leaving the skin mostly untouched and only eating the flesh, but not the seeds (Everett and Everett 2019). Sometimes fruit still on the tree is pecked off, for instance dogwood (*Cornus*) berries in autumn (Snow and Snow 1988).

Sapsucking

One study mentions that Green Woodpeckers drill small holes in tree trunks, mainly pine, fir and yew, in spring to obtain sap (Turček 1954). This method of foraging is known as 'ringing' or 'girdling' and the holes made are called 'wells'. Though some other woodpecker species feed in this way (for example, in Europe most notably Great Spotted Woodpeckers, and in North America four species in the *Sphyrapicus* genus, aptly named 'sapsuckers'), it does not appear to be common behaviour for Green Woodpeckers (Gorman 2015).

Food fed to young

Nestlings are fed largely on a diet of ants, often pupae, but not exclusively. A study of a nest in Romania found that ten species of ant were fed to nestlings. The seven chicks received increasing amounts of food as they grew and consumed an estimated 1.5 million ants, in various forms, before they finally fledged (Korodi Gal 1975). Parents do not usually carry food to the nest in the

FIGURE 15.3 Food is not passed but thrust directly into the bill and throats of Green Woodpecker nestlings by their parents. Emilia-Romagna, Italy, May 2017 (ET).

bill, but rather store it in a crop-like pouch before regurgitating it directly into the throats of their young. Although Green Woodpeckers feed their brood all day long, they generally visit them less often than most other woodpecker species. It has been observed, for instance, that Great Spotted Woodpeckers take food to their nestlings some 6–10 times more often (Sielmann 1961). It is likely that so few visits are possible because more food can be carried in a crop than in a bill (Tutt 1956).

Finding food

Abundant food resources, such as ant colonies, seem to be remembered and visited repeatedly. But how do Green Woodpeckers find these places? How do they locate their food? Which senses do they employ? Do they simply rely upon sight or are non-visual cues from senses such as hearing (auditory), smell (olfactory), touch (tactile) also used?

Vision

Sight is presumed to be the primary sense that most birds employ when searching for food. Even in those birds that are known to use hearing, olfaction and touch to locate food, such as some shorebirds (Scolopacidae) and petrels (Procellariidae), vision is still the main sense they employ in their daily activities.

FIGURE 15.4 A pasture dotted with the mounds of Yellow Meadow Ant, one of the favourite prey species of Green Woodpeckers. It is thought that such sites are found visually. Börzsöny Hills, Hungary, February 2022 (GG).

FIGURE 15.5 A Green Woodpecker uses its long tongue to search for food in a rotting tree. Novo Yankovo, Bulgaria, December 2020 (DG).

Indeed, it is known that much of a bird's brain is devoted to the analysis of information gained from sight (Martin 2021). In addition, it is thought that Green Woodpeckers rely on structural signs, which they use as landmarks when selecting foraging areas, and in doing so they may overlook other areas that are rich in ants (Rolstad et al. 2000).

Touch

Besides locating food visually, woodpeckers also use their tongues to detect prey when it is out of sight (Bock 1999). This long, tactile organ is packed with receptors called Herbst corpuscles that react to vibrations (Martin 2021). Is the tip of the bill also sensitive? Can it 'feel' ants and other insects when the bird cannot actually see them? Bill-tip organs with clusters of touch sensors are known to exist in some other bird families, such as parrots (Psittacidae) and shorebirds, but no evidence for their presence in woodpeckers currently exists.

Hearing

It is also possible that concealed invertebrates are located by sound detection as Green Woodpeckers often probe and dig into substrates when no surface prey activity is evident. Perhaps the high-frequency sounds and vibrations produced by invertebrates can be heard? As mentioned later in this chapter,

FIGURE 15.6 A Green Woodpecker prodding its bill deep into the ground in search of food. Toulon, France, May 2019 (JMB).

Green Woodpeckers will bore holes in beehives to get at the insects inside. It is believed that the humming and vibrating sounds made by the hidden bees may attract the woodpeckers (Cramp 1985). The truth is, we do not fully understand how their senses function, though it is known that the different senses interact, and that non-visual information gathering is often essential for many birds when they look for food (Martin 2021). Ultimately, it is likely that a combination of senses, both 'near-receiving' like touch and 'far-receiving' like vision, is employed by foraging Green Woodpeckers.

Foraging niche

Feeding can begin soon after sunrise and usually finishes well before dusk. Most foraging is done on the ground, with short-grazed and mown grasslands preferred. Areas with tall or dense vegetation, or heavily compacted soil, are avoided (Raqué and Ruge 1999; Alder and Marsden 2010). Birds will also feed in trees, bushes, dry streambeds, cliffs, stone quarries, brick walls and roofs, and search for insects in cowpats, fungi, spider-webs and leaf-litter. Although most other woodpecker species in their range will, to varying degrees, feed on the ground, Green Woodpeckers have become the specialists of the grassland habitat niche.

Partitioning

How similar species use an environment differently to enable them to coexist is termed 'niche partitioning'. How they use the resources within an environment to avoid competition is known as 'resource partitioning'. These methods of separation are often very subtle. For example, although they often occur together, the Green Woodpecker has a more terrestrial feeding niche than its close relative the Grey-headed Woodpecker and is more often found in open areas away from forests proper (Winkler et al. 1995; Gorman 2004). Though they may be slight, morphological factors, such as body size, and bill and tongue lengths, also influence where the different species forage and what they consume.

Foraging methods

Unlike some other bird families, which use their feet and claws to find food – gamebirds scratching at soil, thrushes sweeping away leaf-litter or raptors clutching prey, and so on – woodpeckers only use their bill. Foraging Green

FIGURE 15.7 This female's bill is coated in snow and soil after digging for food. Kocsér, Hungary, January 2011 (RP).

Woodpeckers employ their bill in various ways: as a lever to prise up turf; like a spade to prod and dig into soil, turf, ant mounds and through snow; and to sweep away leaf-litter. They also use the bill to peck and hack into soft or rotten wood and lever off loose bark but seldom riddle trees with large, deep cavities (Gorman 1995). They employ their long sticky-tipped tongue as a probe, inserting it into places too narrow for the bill, such as cracks in the ground, gaps between paving stones and passages inside ant colonies, and to glean prey directly from surfaces.

When a rich source of ants is found in the ground, Green Woodpeckers will excavate small pits to reach them (see Chapter 14, Tracks and Signs). If these feeding sites prove productive, several holes are made, each visited repeatedly in rotation over a period of several hours per day over several days, and even weeks (Löhrl 1977). The birds will enter them with their head and upper body but rarely the whole body, rather preferring to insert their tongues to catch prey. Green Woodpeckers possibly visit feeding holes in rotation not to avoid over-exploiting them, but because they can only tolerate swarming and biting ants for brief periods of time. The fact that they often scratch themselves with their feet and bill, ruffle their feathers and even move away to preen during feeding sessions, seems to indicate this.

FIGURES 15.8a and 15.8b Green Woodpeckers will dig through snow to reach prey in the soil beneath. Budapest, Hungary, December 2012 (GG).

Foraging through snow

Green Woodpeckers are sensitive to prolonged periods of snow, as it affects their ability to access food. Yet when an ant colony is buried by snow, birds will swish their bill from side to side to expose the insects beneath. Though ant colonies can be essential food sources in winter, they are rarely over-exploited, although individual woodpeckers may visit profitable sites repeatedly. This is partly because Green Woodpeckers do not normally dig more than 60 cm into ant nests, and in winter most *Formica* worker ants reside out of their reach at depths of at least 100 cm (de Bruyn et al. 1972). Despite this, when an ant colony has been repeatedly exploited in winter, the insects may relocate the following spring, though sometimes this can involve a journey of only a few metres.

When ants are in short supply or difficult to access, other insect food is turned to. As already mentioned, Green Woodpeckers can be resourceful feeders. For instance, an examination of droppings showed that some Green Woodpeckers in the Netherlands had eaten considerable amounts (31% of their diet) of Birch Catkin Bugs *Kleidocerys resedae* in winter (Klosters et al. 2014).

Foraging on walls and buildings

In times of heavy snow or when the upper layer of soil is frozen, feeding on the ground can be severely hindered, as in those conditions ants often become dormant and difficult to retrieve (de Bruyn et al. 1972). To address this, Green Woodpeckers can be resourceful and visit vertical surfaces such as cliffs, rock walls, stone quarries and buildings to forage (Baier 1973; Löhrl 1977; Blume 1996). In a study by the author in Hungary, individuals foraging on walls in abandoned and active stone quarries, limestone cliffs, and on the walls and rooftops of houses, apartment blocks, hotels and a ruined military base were documented (Gorman and Alder 2022). It was found that the woodpeckers visited these places mainly during periods of ground snow cover because they remained largely free of snow. In addition, most birds visited in afternoon hours which seemed to indicate that warmth by the sun was important as invertebrates would presumably be more active and accessible then. It was surmised that birds foraged at these locations due to prey becoming increasingly more difficult to obtain at their usual grassy foraging sites during harsh conditions.

On buildings, Green Woodpeckers explore brick and stone walls, but also look into gutters and beneath eaves, and inspect thatch. Hosking (2011)

FIGURE 15.9 A Green Woodpecker searching for prey in a thatched roof. Suffolk, England, October 2009 (DH).

reported on a bird that made regular and prolonged visits to a newly thatched roof, probing through the wire netting into the straw, mainly on the south-facing end of the roof. Foraging on houses is likely to be much more common than reported, as the skittishness of Green Woodpeckers probably means that birds quickly flee when people approach. The reason that this behaviour is not as frequently observed in Britain as it is on the continent is most likely because of the relatively mild winters there.

Changing feeding sites in harsh weather is not unusual in the *Picus* genus. Grey-headed Woodpeckers also move their feeding locations, and hence prey sought, when deep snow or frozen ground impedes them (Rolstad and Rolstad 1995; Edenius at al. 1999; Gorman 2020a). Since it is a mainly terrestrial feeder, the shift to foraging on surfaces other than the ground by the Green Woodpecker is presumably an adaptive response to variations in prey availability and accessibility. In some regions, surfaces such as cliffs and those in stone quarries and on buildings may be vital in periods of deep snow to sustain this species, which is vulnerable to harsh weather (Rolstad et al. 2000).

FIGURE 15.10 Although ant specialists, Green Woodpeckers will exploit other food resources. Here a juvenile forages for invertebrates on a shellfish-encrusted wooden groyne support. Shellness, Isle of Sheppey, Kent, England, August 2022 (NU).

FIGURE 15.11 A beekeeper has placed netting over one of his hives in an attempt to deter a visiting woodpecker. Nógrád, Hungary (GG).

FIGURE 15.12 A male Green Woodpecker leans his head back to swallow water taken from a pool. Emilia-Romagna, Italy, August 2020 (ET).

Breaking into beehives

As mentioned in Chapter 8, Green Woodpeckers also show their adaptive side when they visit honeybee hives which they open up to get at the insects, and apparently also the honeycomb, inside. It is thought that this foraging behaviour is learned, and localised, as not all Green Woodpeckers take advantage of this easy food resource. Hives are usually attacked in winter, when temperatures drop, presumably because other food has become scarce or difficult to obtain. There are also records of Green Woodpeckers reacting to harsher feeding conditions by taking wild solitary bees from their nests in plant stems (Poma 1999).

Drinking

Though not often observed, Green Woodpeckers do indeed drink, taking water from natural pools, puddles, garden ponds and hollows in trees. Drinking is often done together with bathing (see Chapter 10, Behaviour).

Chapter 16
Relationships

Interactions between animals can be divided into two basic types: intraspecific and interspecific. We can define these terms here simply as the relationships Green Woodpeckers have with others of their own species (intraspecific) and those they have with other wildlife (interspecific). When interactions involve conflict – and antagonistic exchanges between individual Green Woodpeckers and between Green Woodpeckers and other animals do occur – they ultimately result in an overall balance within wildlife communities being established.

Intraspecific

Interactions are probably more common between different Green Woodpeckers than they are between individual Green Woodpeckers and other species. The most common forms of intraspecific relationships concern those between a mated pair and between parents and their young (see Chapter 12, Breeding). Disputes and aggressive interactions usually involve two males and mostly occur in early spring when territories are being occupied, calling is at its peak and pairs are being formed. They become less frequent once nesting is underway. Individuals seldom feed close to one another. When birds do meet at the same food resource, an aggressive encounter usually ensues. Actual physical contact is unusual in such encounters (see Chapter 10, Behaviour).

Interspecific

Despite regularly occurring alongside several other woodpecker species, Green Woodpeckers only occasionally interact with their sympatric relatives and overly aggressive encounters are uncommon. Although there is some competition for food resources, their specific diet of ground-dwelling ants

FIGURE 16.1 Two female Green Woodpeckers aggressively confront each other. Łazienki Park, Warsaw, Poland, March 2013 (MS).

living in open grasslands means that Green Woodpeckers are generally free from competition. However, a trade-off for reduced competition is that Green Woodpeckers, unlike their relatives that inhabit closed forests, probably face an increased risk of predation when foraging in the open (Elchuk and Wiebe 2002). In some circumstances, they neither ignore nor aggressively engage with other woodpecker species. On the contrary, they occasionally breed with them. For more on hybridisation between this species and other members of the *Picus* genus see Chapter 5, Relatives.

Green and Grey interactions

Green and Grey-headed Woodpeckers are close relatives and sympatric in some areas of continental Europe and Scandinavia. Furthermore, there is overlap in the habitats they use, and hence competition for food does exist between the two, as it does for suitable nesting and roosting trees (Blume 1996). But to what extent? Does the larger size, and hence ultimately dominance, of Green Woodpeckers lead to pressures that result in the exclusion or declines in Grey-headed populations where the two species coincide? In general, the two tolerate each other (perhaps more correctly, we might say they do not seriously conflict with each other). However, when the habitat in which they coexist is poor quality, when food resources are depleted, when the area is small, what happens then? Do they respond by becoming less accepting of each other? Do disputes increase? It has been suggested that the smaller Grey-headed Woodpecker benefits when Green Woodpeckers are absent (Glutz von Blotzheim and Bauer 1994). For Green Woodpeckers to thrive, they need fairly open forests or woodlands, some larger, mature deciduous trees and certain kinds of ants to be abundant. In the case of Grey-headed Woodpeckers, these factors are less important. They are more arboreal, more associated with closed forests than open ones, frequently found in forest interiors, and are less dependent on terrestrial ants as food (Blume 1996; Winkler et al. 1995; Gorman 2004). In a nutshell, when they co-occur, their foraging niches differ. The two species subtly use different parts of the habitat and eat different things, and this alleviates competition (see Chapter 15, Foraging and Food).

Green and Black interactions

In some regions, the relationship between Green Woodpeckers and Black Woodpeckers develops in winter when deep snow cover can force the former to turn to more arboreal prey such as carpenter ants, which have their colonies in tree stumps. These large ants are the preferred prey of Black Woodpeckers (Gorman 2010). Yet even at such times confrontations are usually averted because of a simple order of dominance based on size. Green Woodpeckers are submissive to the larger Black Woodpecker but not to any other woodpecker species. Competition for nesting and roosting sites can also result in interactions between the species (Blume 1996). These two woodpeckers (and also Grey-headed) will, nevertheless, often nest relatively close to each other, even within 100 metres (Stenberg and Hogstad 1992).

Green and Wryneck interactions

In continental Europe, Green Woodpeckers are also sympatric with the Wryneck. They often occupy similar habitats and even feed upon the same kinds of terrestrial ants that live in open areas. When a researcher in Britain analysed Wryneck droppings, he noticed that the contents, mostly fragments of *Lasius* ant workers and cocoons, were very much like those of Green Woodpeckers he had previously examined (Speight 1974). Despite this, competition and interactions in general between these two picid species are uncommon (Gorman 2022). When the two species do meet, for example at an ant nest, the much smaller Wryneck always gives way.

FIGURE 16.2 In many areas Wrynecks consume the same ant species as Green Woodpeckers and also catch them in a similar way, by using their long sticky tongue. Here a few of the insects have avoided being eaten by crawling on top of the bird's head. Fejér County, Hungary, June 2012 (AK).

Green and Great Spotted interactions

These two species are sympatric across much of Europe, yet even where they are both common interactions between them are, perhaps surprisingly, infrequent. An extraordinary observation in Britain concerned a female Green Woodpecker sharing a nesting cavity, complete with clutch, with a male Great Spotted Woodpecker (Hoffman 1951). Both birds were seen entering and leaving the hole. The observer felt the mixed pair had mated but it is perhaps more likely that hole competition resulted in this unusual couple. Ultimately, to truly understand how various woodpeckers can co-exist and be successful, the ecologies of all the species involved need to be thoroughly understood.

Other birds

When encountering other (non-woodpecker) birds at feeding sites, such as a pile of windfall fruit, Green Woodpeckers are normally dominant over birds such as Eurasian Blackbird *Turdus merula* and Fieldfare *Turdus pilaris*, but usually not over corvids such as Eurasian Jay *Garrulus glandarius*. As mentioned in Chapter 10, Green Woodpeckers cannot reasonably be described as social birds. They have no symbiotic relationships, yet instances of commensalism can occur. An example observed in England in winter involved a Woodpigeon *Columba palumbus* which remained in close attendance to a Green Woodpecker that was energetically stabbing apples, in order to pick up the small fragments that were being scattered (Everett and Everett 2019). Another winter observation from England involved a Green Woodpecker digging up and feeding on ants on a garden lawn. As it moved from one spot to another, it was followed by four Common Starlings which collected the stray ants that the woodpecker had missed (King et al. 1973). Another curious relationship was documented in France in May 2014 (Perrot 2015). A pair of Little Owls nested in a former Green Woodpecker nesting cavity, in a plane tree, that had two entrances, after having nested some 40 metres away the previous year in a natural cavity in an apple tree. Although the woodpeckers did not breed in the plane tree in 2014, they roosted there each night, seemingly sharing the same cavity as the owls. No physical aggression between the two species was recorded. The observer concluded that either the owls bred at the bottom of the chamber,

FIGURE 16.3 A male leaves a freshly made nesting cavity. In subsequent years this hole may be used by the woodpeckers themselves again, or by a range of other birds, mammals and insects. Nógrád, Hungary, April 2022 (GG).

while the woodpeckers slept clinging to the chamber wall above, or that there were two separate chambers inside the tree that could be accessed from the two entrances and that the owls bred in one and the woodpeckers slept in the other.

Cavity usurpation

Another example of interspecific behaviour occurs when another animal attempts to occupy an active woodpecker cavity. The responses of Green Woodpeckers to this vary, with some pairs or individuals being assertive while others are more passive. Much depends upon which species is attempting to usurp their cavity. Once again, size matters and Green Woodpeckers usually dominate smaller bird species which seek to use their cavities, and which tend to give way (Keicher 2007). Nevertheless, some Green Woodpecker pairs seem to have less conviction than others and surrender their cavities when the opposition, such as Common Starlings, is persistent.

Keystone species

Ecologists and naturalists use the term keystone species to represent an animal that plays a crucial role in the ecosystem where it lives by helping the ecosystem retain its diversity, form and functionality (Mikusiński and Angelstam 1997). Keystones have a disproportionately significant effect on their environment in relation to their abundance, with their presence affecting many other organisms. They modify habitats in ways that influence the other animals that use the habitat, helping determine their abundance and condition. If a keystone species disappears, the environment changes, even though that species was just one small part of it. Green Woodpeckers are very much part of the woodland and grassland communities they live in and can be regarded as keystones mainly because (like other woodpeckers) they excavate cavities that are important, sometimes indispensable, for many other species of birds, mammals and invertebrates which cannot do so. The list of secondary cavity users known to use their holes includes numerous birds, mammals and insects (see Chapter 13, Cavities).

Predators and prey

One of the most common ways in which different species engage and interact with one another is in a predator–prey relationship. Naturally, Green Woodpeckers are taken by some predators, but they do not form the staple diet of any particular species, whether avian, reptilian or mammalian. The most important predator–prey relationship that Green Woodpeckers are involved in actually does not concern these birds as prey but as predators, that is, of ants, their preferred and unwilling food resource (for more on this see Chapter 15, Foraging and Food).

Avian predators

Green Woodpeckers are occasionally taken by owls and raptors such as hawks, harriers and larger falcons. It is not unusual to find their remains in the nests of Peregrines *Falco peregrinus* (Rockenbauch 2002), and in Central Italy they are apparently frequently taken by Lanner Falcons *Falco biarmicus* (Morimando and Pezzo 1997). They are also preyed upon by Northern Goshawks *Accipiter gentilis* (Opdam et al. 1977) and Eurasian Sparrowhawks *Accipiter nisus*, but usually females of the latter as males are too small to tackle them. Perhaps

FIGURE 16.4 In open environments a Green Woodpecker in flight is no match for a Peregrine, as this unfortunate juvenile found out. Christ Church, Cheltenham, England, July 2011 (DP).

FIGURE 16.5 Aware of a potential threat, this male Green Woodpecker raises his red crown feathers and intently surveys the scene. Toulon, France, March 2020 (JMB).

more surprisingly, Green Woodpeckers have also been recorded as prey of Golden Eagles *Aquila chrysaetos* (Clouet et al. 2015). Their remains are also sometimes found in the pellets of Tawny Owls *Strix aluco* (Obuch 2011) and Eagle Owls *Bubo bubo* (Sándor and Ionescu 2009).

Mammalian predators

Mammals such as martens, stoats and weasels occasionally prey on Green Woodpeckers. In this regard, it is impossible not to mention the extraordinary incident that took place in Essex, England in March 2015 when a Green Woodpecker was seen taking flight with a Common Weasel *Mustela nivalis* clinging to its back. Some scepticism ensued after photographs of the event went viral on social media, but the observation was corroborated. Indeed, though weasels are smaller than Green Woodpeckers they are fearless predators

FIGURE 16.6 A juvenile Green Woodpecker hiding in the cover of a leafy tree, its green plumage providing a degree of camouflage. Czech Republic, August 2013 (TG).

that usually hunt for prey on the ground, where the woodpecker in question was foraging. Also in Britain, the non-native Grey Squirrel will take woodpecker eggs (Glue and Boswell 1994). Newson et al. (2010) found that where these invasive mammals were abundant, they contributed to nest failure rates and lower levels of population growth in several woodland bird species, including Green Woodpeckers.

Anti-predator strategies

To avert being predated, Green Woodpeckers use a variety of tactics. Often, when first sensing danger, they remain silent and still, freezing on a trunk or branch. Or they may make raucous alarm calls from within cover or simply hide in silence. Some individuals seem to be bolder than others and will mob or fly at and chase off smaller raptors such as Eurasian Sparrowhawks while making loud calls (Capps 2002). It is also thought that Green Woodpeckers may avoid foraging in taller, denser grassland, although they often host many ant colonies, due to an increased risk of predation in such places (Alder and Marsden 2010).

Chapter 17

Folklore, Mythology and Symbolism

The Green Woodpeckers flying up and down
With wings of mellow green and speckled crown
She bore a hole in trees with crawking noise
And pelted down and often catched by boys

Extract from 'The Green Woodpecker's Nest'
by John Clare (1832–7)

An intriguing and rich body of fables, myths and superstitions has developed around the Green Woodpecker. Across Europe, people have considered the bird magical, often a good omen but sometimes a portent of bad luck. In ancient times it was deemed an oracle, a totem, a weather forecaster and was variously associated with war, agriculture, desire and fertility. In some interpretations of Babylonian, Greek and Roman legends, the woodpecker is a bird with supernatural powers, but sometimes it has appeared in the guise of a king and even a god. Many of the traits that the ancients bestowed upon the Green Woodpecker endured and subsequently became entwined in the folklores and mythologies of cultures that followed.

'Mesopotamian Woodpecker' revisited

The Green Woodpecker's symbolic significance goes back a long way. The Babylonian Empire arose around 4,000 years ago in Mesopotamia, within the floodplains of the Tigris and Euphrates rivers in what became known as the Fertile Crescent. In this culture a mysterious deity, the Axe of Ishtar, a goddess that embodied love, desire, fertility and rebirth, and who was associated with Venus, was represented as a green-coloured woodpecker

FIGURE 17.1 *Picus viridis*, 1876. From *Onze vogels in huis en tuin (Our birds in home and garden), 1869–1876* by John Gerrard Keuleman (WikiCommons).

(Clare Lees 2002). Subsequently, the myth of this sacred woodpecker went westwards from Babylon.

Ancient and Classical Greece

Greek mythology is alive with woodpeckers, and they are often green. In Crete, the mythical woodpecker-like god Picus was associated with Zeus, the ultimate Greek god. This is particularly puzzling as there are no woodpeckers of any kind on Crete today (with the exception of the migratory Wryneck) and it is most likely that there were none in the past, either. In some accounts Picus and Zeus were the same. In another Greek legend a woodpecker (it is not clear whether Green or another species) ruled as king, before Zeus usurped him. In his *Metamorphoses* (written sometime between 100 and 300 CE and not to be confused with Ovid's work of the same name), Antoninus Liberalis relates how King Celeus of Eleusis tried to steal honey intended for the infant Zeus from bees in a cave on Mount Ida. Rather than simply slaying Celeus and his cohorts in the sacred place of his birth, Zeus transformed them into birds, Celeus becoming a Green Woodpecker. In another, typically rather convoluted fable involving abduction, rape, murder and cannibalism, Zeus punished several characters who had apparently insulted him by transforming them into birds. Polytechnos, who was among other trades a carpenter (appropriately), was changed into a Green Woodpecker in retribution (Clare Lees 2002).

The Piceni

The Piceni people lived in Picenum on the coastal plain between the Apennine mountains and the Adriatic Sea (roughly today's Marche and Abruzzi regions in Italy) between the ninth and third centuries BCE. Also referred to as the Picentes and Picentini, very little is actually known about this pre-Roman tribe. Legend has it that they were people 'of the woodpecker', with the bird playing a central role in their culture and beliefs. The ancestors of the Piceni are said to have journeyed westwards across the Adriatic in search of a new home. When they disembarked, a Green Woodpecker settled upon one of their banners, and this was taken to be an omen that they should follow the bird. When they arrived at the banks of the River Tronto, the woodpecker landed in a tree and began to make a hole. This was interpreted as a sign to settle at that place, and so they did, adopting the woodpecker as their totem and sacred symbol. One version of this myth states that the woodpecker that led them was *Picus*, a woodland deity that appears in later Roman mythology, and that from him the tribe took their name (Clare Lees 2002).

Marche's woodpeckers

The Italian town of Ascoli Piceno nestles in the Marche region. It is said to lie on the site where the Piceni settled after following their woodpecker and this pagan legend, despite the purging efforts of Christianity, remains to this day. Indeed, a woodpecker adorns the emblem of Marche, and one of the local football clubs is called Ascoli Picchio F.C., the 'Ascoli Woodpeckers'. Marche's picid past is also celebrated every Whitsun in the hill town of Monterubbiano in a pageant called the *Scio la Pica*, the 'Woodpecker Chase', when a woodpecker (though today another bird is used) attached to a cherry branch is paraded through the streets.

Ancient and Classical Rome

Roman culture and beliefs were profoundly influenced by the various peoples who came before them, including Greeks, of course, but also those who lived on the Italian peninsula, such as the Piceni (see above). Some of the Roman counterparts of the Greek gods inherited their woodpecker associations, Jupiter following his predecessor Zeus. and Silvanus, the god of forests, following Pan. Mars, like Ares, was the god of war but also of agriculture and the woodpecker was allied to him in both roles. Ovid called the woodpecker 'the bird of Mars' and Plutarch wrote 'esteemed holy to the god Mars; the woodpecker the Latins still especially worship and honour' (Clare Lees 2002). Fittingly, another

FIGURE 17.2 A woodpecker graces the emblem of Marche in Italy, recalling the legend of the bird that guided the Piceni to that region (WikiCommons).

FIGURE 17.3
Luca Giordano's *Picus and Circe*, *c*.1530–1540. The man with woodpecker wings tries to flee from the spurned sorceress (WikiCommons).

woodland deity was called Picus whose many talents included being able to metamorphose. In a particularly colourful legend, Picus the woodpecker was originally a human, the king of Ausonia, founder of Alba Longa. The story goes that while he was on a hunting trip, he was watched by the sorceress Circe, who fell in love with him. Using her charms, she separated Picus from his friends and proclaimed her love for him. When he rejected her advances saying he would always be faithful to his wife, the furious Circe changed him into a woodpecker.

Ornithomancy

The ritual of foretelling events from the calls and behaviour of birds, known as ornithomancy, was practised in ancient Greece and by various tribes on the Italian peninsula. For example, in the fifth century BCE the Aequi people, who lived in the forests of the Apennine Mountains above Rome, believed that messages from the god Mars were conveyed to them by a celestial woodpecker. In some accounts the woodpecker does not appear to be a mere messenger but an actual zoo-anthropic manifestation of Mars himself. In Rome itself, flying birds and calling birds were studied for omens, and woodpeckers fell into both camps. It was believed, for instance, that the outcome of future events could be deduced from observing the direction in which a woodpecker flew. In one particularly bizarre legend, a woodpecker was said to have landed on the head of Aelius Tubero, the city praetor, as he was passing judgement on a legal matter in the forum. The woodpecker was captured, and the extraordinary event debated with the conclusion being that Rome would be in danger if the bird were released, but that tragedy awaited the praetor if it was harmed. Tubero promptly did his duty and killed the poor woodpecker (some accounts say he bit its head off) so that Rome would prevail, but as predicted, the loyal Roman shortly came to a gruesome end. The Roman woodland deity Picus, who we met briefly above, was said to be able to predict the will of the Gods by observing signs in the sky. Picus lived just outside Rome on Aventine Hill, perched on a wooden column in a dark oak grove, where he was often joined by Faunus, another nature god. Finally, the legend of how a she-wolf protected the twins Romulus and Remus, the founders of Rome, is well known, but what is less commonly known is that, according to Ovid, a woodpecker also helped guard the boys. It is likely that the avian foster parent was a Green Woodpecker, given the cult that surrounded the bird, but as is often the case with such myths and legends, the exact identity of the species is unclear.

A wealth of woodpecker names

The *Icones animalium* was published in 1560 by Conrad Gesner. This compilation of some 150 bird species known to him in England includes *Picus viridis*. The English name 'Green Woodpecker' can, however, be traced to well before that, to the eighth-century Anglo-Saxon Chronicle (Bircham 2007). But back then, everyday English folk would almost certainly not have used the name we now do for this bird; rather they would have had a local one, and many existed, usually differing from county to county. Some, like 'tapper', 'wood hack', 'wood knacker', 'wood spack', 'wood speck', 'wood speight' (also simply 'speight'), 'wood spite' and 'wood sucker', were generic terms used for all woodpeckers (Lockwood 1984). The following names all specifically referred to the Green Woodpecker: 'woodwall' in Somerset (apparently used for Great Spotted Woodpecker in Hampshire); 'wood hack' and 'greenpeek' in Lincolnshire; 'hickwall' (derived from the fifteenth-century 'hykwale') and 'hew-hole' in Northumberland, with variants 'heyhoe', 'heighaw', 'heigh hold', 'high hoe', 'yough-all', 'highaw', 'ickwall', 'etwall', 'eckwall', 'eckle', 'ickle' and 'nickle'; in Northamptonshire, Oxfordshire and Warwickshire 'hickel'; in Cornwall 'hood awl', and elsewhere 'awl bird'; and in Gloucestershire 'greenile'. In many regions, the Green Woodpecker had names that associated it with coming rain, the simplest being 'rain-bird' and 'rain-fowl', but also 'rain-pie', 'wet-bird', 'weather-hatcher', 'weather-cock' and 'storm-cock'. Similarly, an old local Welsh name was *Caseg y drycin* ('storm mare'). For more on this belief, see below. Obviously, alternative old and local names, as well as modern ones, also abound in other languages wherever Green Woodpeckers are found.

FIGURE 17.4 The Green Woodpecker by Thomas Pennant. From *The natural history of British birds, or, A selection of the most rare, beautiful and interesting birds which inhabit this country – the descriptions from the Systema naturae of Linnaeus* (WikiCommons).

The laughing bird

The well-known folk name 'yaffle', and similar ones from around England such as 'yaffler', 'yaffingale', 'yappingale', 'yelpingale', 'yippingale', 'yockle' and 'yuckle', are clearly onomatopoeic, all inspired by the bird's 'song', which for many people reminded them of human laughter. A West Yorkshire name for the bird, 'heffald', is also apparently a derivative of 'laugh'. In the southern counties of Kent, Surrey and Sussex, 'galley bird' evidently also refers to its loud call. In Gloucestershire, it was called 'laughing Betsy'. In Shropshire, they kept it simple, naming the Green Woodpecker the 'laughing bird' (Lockwood 1984).

The weather woodpecker

Many superstitions and sayings from all over Europe associate the Green Woodpecker with the weather. Wherever it was found across Scandinavia, the Green Woodpecker was said to be a reliable weather forecaster and was therefore respected (Clare Lees 2002). In Bohemia, in today's Czech Republic, it was believed that if one called during winter, then that season would be longer than usual, and spring would arrive late, and hence farmers took note and planned accordingly. Of all weather events, the one that is most commonly associated with Green Woodpeckers is rain.

The rainbird

The Romans named the Green Woodpecker the *pluviae aves*, the 'rainbird'. It was attributed with being an avian weather forecaster, and this belief later persisted across Italy where folk would plan their days in response to hearing the woodpecker call. This belief also followed the Romans as they marched across Europe and found its way into the folklore and mythologies of the peoples they conquered. Then again, in some regions the Green Woodpecker's relationship with precipitation may have already existed, as in Britain where the Druids are believed to have regarded it as a bringer of rain (as mentioned above, several old names for the bird in England refer to this association). In German-speaking countries, too, there are many folk names that connect woodpeckers and rain, though not always exclusively Green Woodpeckers. For example, *Giesser* (pourer) and *Gießvogel* (pouring bird) are quite common names for the Green Woodpecker, but in Switzerland they refer to the Black Woodpecker and in Austria sometimes the Wryneck. In parts of France the species is still known as *pleu-pleu* (rain-rain). Also in France, in the days when

FIGURE 17.5 A male Green Woodpecker stops for a drink by a garden pond. Proof, if it was needed, that the belief that woodpeckers were doomed by God to never drink from standing water is indeed a myth. Norfolk, England, May 2008 (NB).

watermills were numerous, it was called *l'avocat, ou le procureur du meunier* (the miller's lawyer), because its calls were believed to attract rain which, besides quenching its thirst, also filled the streams and drove the mills, much to the satisfaction of millers who were thus always glad to see these birds (Tate 2007).

The time when a Green Woodpecker called was also often significant. In Sweden hearing one in summer meant imminent rain, but when it called in winter mild weather was on the way (Svanberg 2013). Similarly, a rural French saying asserts '*Si en juillet le pivert crie, il annonce la pluie*' (If the Green Woodpecker calls in July, it is announcing rain). In more modern times, an obsolete scientific name given to a previously claimed subspecies found only in Britain was *Picus pluvius*, again following the old superstition that its 'laughter' was a portent of an approaching downpour. A fable that is told across rural Europe seeks to explain, from a Christian creationist perspective, how the woodpecker became the 'rainbird'. It runs something like this: when God created the world, he wanted to provide springs, ponds, lakes and rivers for all to drink from, so he invited those birds with strong bills to dig holes which would then fill with water. But the slothful Green Woodpecker refused, and simply flew around doing his own business. God was not pleased and punished the bird for not helping to dig, condemning it to forever peck wood to find water. And that is why, when a woodpecker is thirsty, it is forced to look up towards heaven and plead with repeated calls for rain so that it can lick moisture from leaves and branches (Tate 2007).

Farming and fertility

As detailed in Chapter 2, the Green Woodpecker uses its long, sturdy bill to probe and dig into ground for invertebrate prey. We have also heard how a cult arose around the Green Woodpecker in Babylonia, the region where agriculture is believed to have first been practised in the so-called Fertile Crescent. Consequently, a connection between the Green Woodpecker and

FIGURE 17.6 A female Green Woodpecker digs into the ground with its bill in search of prey. It is said that in ancient times as agriculture developed, people observing this behaviour associated it with the tilling of the soil and the sowing of seeds. Toulon, France, April 2020 (JMB).

FIGURE 17.7 Detail from the Beatus page of the thirteenth-century Alphonso Psalter with a Green Woodpecker and a Kingfisher (WikiCommons).

farming and fertility has been proposed. In particular the woodpecker's bill has been depicted as a tool used for working the land and especially as a plough, an implement believed to have originated in that region (Gorman 2017).

Good luck, bad luck

The 'laughing' call of the Green Woodpecker has not only inspired many local folk names but has been interpreted as mocking or as a warning. In some corners of Europe hearing the call come from the right meant good luck, but if the call came from the left, then something unlucky would result. Some also believed that if the woodpecker laughed once or twice, all was well, but hearing it call three times meant bad luck. Similarly, a Green Woodpecker seen flying from left to right was a good omen, but not one going the other way. These superstitions probably have their origins in Roman divination rituals (see above), where the direction of the woodpecker's calls and flight were regarded as prophetic.

The woodpecker's magic herb

In continental Europe a particularly strange myth involved the Green Woodpecker and a mysterious herb with magical, strength-giving properties (Tate 2007). Sometimes referred to as 'spring-wort', 'spring-wurzel' or 'moonwort', it has not been reliably identified as any known species – though Maidenhair Fern, St John's-wort and Solomon's Seal, as well as various peonies and spurges, have all been put forward. Whatever it was, this plant (which was sometimes referred to as a root) was highly coveted but could only be obtained with the help of the woodpecker, who knew where to find it, but guarded it vehemently. In parts of France, it was said that when the Green Woodpecker laughed it was mocking those who roamed the countryside trying in vain to find this magic herb. It was believed that the woodpecker was able to peck into the hard wood of trees because it had sharpened its bill against the plant. Hence, it was believed that rubbing the wondrous herb on the limbs would give a person super strength. In Germany, it was said that if one blocked the woodpecker's nest hole, the bird would bring this herb to unblock it. Thus, folk thought it could also be used to open locks without keys. To find where the herb grew, the woodpecker's movements had to be covertly watched, but the bird apparently always took care not to be followed to those secret locations. Anyone who succeeded in finding the herb was advised to collect it under the cover of darkness, otherwise the woodpecker would attack and try to poke the intruder's eyes out. Some legends took this notion further, stating that one could be blinded by simply encountering the woodpecker when trying to find the plant.

FIGURES 17.8, 17.9, 17.10 Three town emblems with Green Woodpeckers. Ganterschwil St Gallen in Switzerland, West-Nieuwland in the Netherlands, and Sitke in Hungary (WikiCommons).

Folk remedies

Along with those of many other animals, the body-parts of Green Woodpeckers have been accredited with having therapeutic properties. Once upon a time in France, eating a fully feathered Green Woodpecker was believed to shield a person from black magic. In the Austrian Tyrol, the flesh of the Green Woodpecker was consumed to treat epilepsy. In medieval Germany white wine blended with ground Green Woodpecker bones was drunk as a cure for kidney and bladder stones, and the woody and faecal mess from the bottom of its nest would alleviate a headache if daubed on the forehead. The twelfth-century German polymath Hildegard von Bingen regarded the species as the strongest woodpecker (she presumably had not encountered a Black Woodpecker), and remarked: 'This bird's nature is clean, its heart simple and without any evil art.' Consequently, she proclaimed: 'A person who is leprous should roast the Green Woodpecker on a fire and eat it often; it will destroy the leprosy.' Another of her medications was an ointment made from the bird's flesh and skin, mixed with vulture and deer fat. When this was used 'frequently to anoint the person's leprosy' they would be cured 'no matter how strong the leprosy is'. Her remedy for arthritis was to 'dry the woodpecker's heart. Set it in gold and silver, as if it were a ring. When you carry it with you, gout will go from you.' As if to prove that none of this was gibberish, von Bingen warned that 'other parts of the woodpecker are not useful for medicine' and that none of her remedies would work 'if the judgement of God does not allow it' (Gorman 2017).

FIGURES 17.11, 17.12, 17.13 Green Woodpeckers have been popular subjects for postage stamps. Here are three, from Moldova, Russia and the Vatican.

FIGURE 17.14 Professor Yaffle, the wise woodpecker from the 1970s children's TV series *Bagpuss* (WikiCommons).

More recent times

In the 1970s, *Bagpuss* was a hugely popular BBC television series for children. Among its somewhat quirky puppet characters was a bespectacled bookend called Professor Yaffle, or to give him his full name Augustus Barclay Yaffle. Apparently modelled upon the polymath Bertrand Russell, the scholarly but somewhat haughty picid became, for a time, the UK's best-known woodpecker.

Certainly, the Green Woodpecker is very appealing, not only to ornithologists but also to non-birders who are often surprised and impressed by its bright colouration, charismatic appearance and behaviour and, not the least, its characteristic 'laughing' call. Long may we hear this wonderful species 'yaffling' at us from its home in the woods.

References

Adamian, M. S. and Klem, D. (1999) *Handbook of the Birds of Armenia*. American University of Armenia, Oakland, CA.

Alder, D. and Marsden, S. (2010) Characteristics of feeding-site selection by breeding Green Woodpeckers *Picus viridis* in a UK agricultural landscape. *Bird Study* 57: 100–7.

Arànega, G. I., Coma, M. Q. and Jimenez-Guri, E. (2020) Divergence between the COI-5P gene regions in the Iberian and European lineages of the Eurasian Green Woodpecker *Picus viridis*. *Revista catalana d'ornitologia* (*Catalan Journal of Ornithology*) 36: 10–20. https://doi.org/10.2436/20.8100.01.16

Baier, E. (1973) Grünspecht (*Picus viridis*) und Grauspecht (*Picus canus*) auf Nahrungs-suche an Hausmauern. *Ornithologische Mitteilungen* 25: 97.

Baker, J. (2016) *Identification of European Non-Passerines*, pp. 423–4. British Trust for Ornithology, Thetford.

Bakken, V., Runde, O. and Tjørve, E. (2006) *Norsk ringmerkingsatlas*. Vol. 2. Stavanger Museum, Stavanger. (In Norwegian with English summary.)

Balmer, D. E., Gillings, S., Caffrey, B. J., Swann, R. L., Downie, I. S. and Fuller, R. J. (2013) *Bird Atlas 2007–11: The breeding and wintering birds of Britain and Ireland*. British Trust for Ornithology, Thetford.

Barker, S. (2021) The value of traditional orchards for birds. *British Birds* 114: 280–91.

Battisti, C. and Dodaro, G. (2016) Mapping bird assemblages in a Mediterranean urban park: Evidence for a shift in dominance towards medium–large body sized species after 26 years. *Belgian Journal Zoology* 146: 81–9. https://doi.org/10.26496/bjz.2016.43

Berger, M. (1990) Eine mögliche Hybridform zwischen Grünspecht und Grauspecht aus Münster. *Natur und Heimat* 50: 109–10.

Bergmann, H.-H. (2018) Weiteres zur Jugendmauser der Spechte: Beispiel Grünspecht *Picus viridis*. *Ornithologischer Anzeiger* 57: 61–4.

Bijlsma, R. G. (2014) Broed- en foerageergedrag van Draaihalzen *Jynx torquilla*. (Breeding and foraging behaviour of Wrynecks *Jynx torquilla*). *Drentse Vogels* 28: 78–100.

Bircham, P. (2007) *A History of Ornithology*. Collins, London.

Bird, D. and Südbeck, P. (2004) Erster Nachweis eines Grünspecht × Grauspecht-Hybriden *Picus viridis* × *P. canus* in Sachsen-Anhalt. *Ornithologische Jahresberichte des Museum Heineanum* 22: 1–3.

BirdLife International (2022a) Species factsheet: *Picus viridis*. http://www.birdlife.org

BirdLife International (2022b) IUCN Red List for birds. http://www.birdlife.org

Blume, D. (1955). Über einige Verhaltensweisen des Grünspechtes in der Fortpflan-zungszeit. *Die Vogelwelt* 76: 193–210.

Blume, D. (1961). Über die Lebensweise einiger Spechtarten (*Dendrocopos major, Picus virirdis, Dryocopus martius*). *Journal of Ornithology* 102 (Suppl.): 1–115.

Blume, D. (1962). Spechtbeobachtungen aus den Jahren 1960 und 1961. *Vogelwelt* 83: 33–48.

Blume, D. (1996) *Schwarzspecht, Grauspecht, Grünspecht*. Neue Brehm-Bücherei, Westarp Wissenschaften.

Blume, D. and Jung, G. (1958) Über die instrumentalen Lautäußerungen bei Schwar-zspecht, Grünspecht, Grauspecht und Buntspecht. *Vogelring* 27: 1–13.

Bock, W. J. (1959) The scansorial foot of the woodpeckers, with comments on the evolution of perching and climbing feet in birds. *American Museum Novitates* No. 1931.

Bock, W. J. (1999) Functional and evolutionary morphology of woodpeckers. *The Ostrich* 70: 23–31. https://doi.org/10.1080/00306525.1999.9639746

Bock, W. J. (2015) Evolutionary morphology of the woodpeckers (Picidae). In H. Winkler and F. Gusenleitner (eds), *Developments in Woodpecker Biology*. Biolo-giezentrum des Oberösterreichischen Landesmuseums, Linz, Austria.

Böhmer, C., Abourachid, A., Wenger, P., Fasquelle, B., Furet, M., Chablat, D. and Chevallereau, C. (2019) Combining precision and power to maximize perfor-mance: A case study of the woodpecker's neck. *44ème congrès de la Société de Biomécanique*, Poitiers, France. hal-02265058. https://doi.org/10.1080/10255842.2020.1713463

Bouvier, J-C., Muller, I., Génard, M., Lescourret, F. and Lavigne, C. (2014) Nest-Site and landscape characteristics affect the distribution of breeding pairs of European Rollers *Coracias garrulus* in an agricultural area of south-eastern France. *Acta Ornithologica* 49: 23–32. https://doi.org/10.3161/000164514X682869

Bradshaw, R. H., Hannon, G. E., and Lister, A. M. (2003) A long-term perspective on ungulate–vegetation interactions. *Forest Ecology and Management* 181(1–2): 267–280. https://doi.org/10.1016/S0378-1127(03)00138-5

Brazaitis G. and Pėtelis K. (2010) The woodpecker guild composition in the forests of central Lithuania. *Acta Biologica Universitatis Daugavpiliensis* 10: 183–8.

Brichetti, P. and Fracasso, G. (2020) *The Birds of Italy*. Vol. 2: *Pteroclidae–Locustellidae*, pp. 141–3. Edizioni Belvedere.

Butler, C. J., Cresswell, W., Gosler, A. and Perrins, C. (2013) The breeding biology of rose-ringed parakeets *Psittacula krameri* in England during a period of rapid population expansion. *Bird Study* 60: 527–32. https://doi.org/10.1080/00063657.2013.836154

Capps, K. (2002) Green Woodpecker chasing Eurasian Sparrowhawk. *British Birds* 95: 24.

Cepák, J., Klvaňa, P., Škopek, J., Schröpfer, L., Jelínek, M., Hořák, D., Formánek, J. and Zárybnický, J. (2008) *Atlas migrace ptáků České a Slovenské republiky* (Czech and Slovak Bird Migration Atlas). Aventinum, Praha. 262–3.

Chatfield, D. G. P. (1970) Abnormally plumaged Green Woodpecker. *British Birds* 63: 429.

Chmielewski, S. (2019) The importance of old, traditionally managed orchards for breeding birds in the agricultural landscape. *Polish Journal of Environmental Studies* 28: 1–8. https://doi.org/10.15244/pjoes/94813

Chodkiewicz, T., Kuczyński, L., Sikora, A., Chylarecki, P., Neubauer, G., Ławicki, Ł. and Stawarczyk, T. (2015) Ocena liczebności populacji ptaków lęgowych w Polsce w latach 2008–2012 (Population estimates of breeding birds in Poland in 2008–2012). *Ornis Polonica* 56: 149–89. (In Polish with English abstract.)

Clare Lees, A. (2002) *The Cult of the Green Bird: The Mythology of the Green Woodpecker.* Scotforth Books, Lancaster, UK.

Clements, J. F., Schulenberg, T. S., Iliff, M. J., Billerman, S. M., Fredericks, T. A., Gerbracht, J. A., Lepage, D., Sullivan, B. L. and Wood, C. L. 2021. *The eBird/ Clements Checklist of Birds of the World: v2021.* https://www.birds.cornell.edu/ clementschecklist/download/

Clouet, M., Gonzalez, L., Laspreses, F. and Rebours, I. (2015) Diet of the Golden Eagle *Aquila chrysaetos* diet during breeding (time) season in the northern Basque country. *Alauda* 83: 1–6.

Cohen, E. (1946) Unrecorded note of Green Woodpecker. *British Birds* 39: 248.

Collette, J. (2008) Les oiseaux du verger en Normandie. *Le Cormoran* 16: 31–57.

Cramp, S. (ed.) (1985) *Handbook of the Birds of Europe, the Middle East and North Africa. The Birds of the Western Palearctic.* Vol. 4. Oxford University Press, Oxford.

Czechowski, P. and Bocheński, M. (2012). A case of hybridization between the Green Woodpecker *Picus viridis* and the Grey-faced Woodpecker *Picus canus. Przegląd Przyrodniczy* 23: 72–4. (In Polish with English summary.)

Danko, S., Darolova, A. and Krištin, A. (eds) (2002) *Rozsirenie vtakov na Slovensku (Bird Distribution in Slovakia).* Veda, Bratislava.

Dean. F. (1949) Green Woodpecker 'drumming' on metal cap of electric pylon. *British Birds* 42: 122–3.

de Bruyn, G. J., Goosen-de Roo, L., Hubregtse-van den Berg, A. I. M. and Feijen, H. R. (1972) Predation of ants by woodpeckers. *Ekologia Polska* 20: 83–91.

del Hoyo, J. and Collar, N. J. (2014) *HBW and BirdLife International Illustrated Checklist of the Birds of the World.* Vol. 1. Lynx Edicions, Barcelona.

Demongin, L. (2016) *Identification Guide to Birds in the Hand,* pp. 208–9. Beauregard-Vendon.

Dmoch, A. 2003. Obserwacja mieszańca dzięciołów zielonosiwego *Picus canus* i zielonego *P. viridis* (Record of a hybrid between Grey-headed Woodpecker *Picus canus* and Green Woodpecker *P. viridis* in Poland). *Notatki ornitologiczne* 44: 273–5.

Dobinson, H. M. and Richards, A. J. (1964) The effects of the severe winter of 1962/63 on birds in Britain. *British Birds* 57: 373–434.

DOF (2020) https://dofbasen.dk (Database of Birdlife Denmark).

Dvorak, M. (2019) *Österreichischer Bericht gemäß. Artikel 12 der Vogelschutzrichtlinie, EG – Berichtszeitraum 2013 bis 2018.* BirdLife Österreich, Wien.

Eaton, M., Aebischer, N., Brown, A. F., Hearn, R., Lock, L., Musgrove, A. J., Noble, D., Stroud, D. and Gregory, R. (2015) Birds of Conservation Concern 4: The population status of birds in the UK, Channel Islands and Isle of Man. *British Birds* 108: 708–46.

EBCC (2015) Pan-European Common Bird Monitoring Scheme. European Bird Census Council. http://www.ebcc.info/index.php?ID=587

Edenius, L., Brodin, T. and Sunesson, P. (1999) Winter behaviour of the Grey-headed Woodpecker *Picus canus* in relation to recent population trends in Sweden. *Ornis Svecica* 9: 65–74. https://doi.org/10.34080/os.v9.22917

Elahi, M., Aliabadian M., Ghasempouri, S. M. and Winkler, H. (2020) Significant divergence and conservatism in the niche evolution of the Eurasian Green Woodpecker complex (Aves, Picidae). *Ecopersia* 8: 109–15.

Elchuk, C. L. and Wiebe, K. L. (2002) Food and predation risk as factors related to foraging locations of Northern Flickers. *Wilson Journal of Ornithology* 114: 349–57. https://doi.org/10.1676/0043-5643(2002)114[0349:FAPRAF]2.0.CO;2

Elts, J., Leito, A., Leivits M., Luigujõe L., Nellis R., Ots M., Tammekänd, I. and Väli, Ü. (2019) Eesti lindude staatus, pesitsusaegne ja talvine arvukus 2013–2017 (Status and numbers of Estonian birds, 2013–2017). *Hirundo* 32: 1–39.

European Environment Agency's Birds Directive 2013–2018 (2018). https://cdr.eionet.europa.eu/help/birds_art12

European Commission (2020) EU Biodiversity Strategy for 2030. Bringing nature back into our lives. Brussels, 20.5.2020 COM (2020) 380 final. https://eur-lex.europa.eu/resource.html?uri=cellar:a3c806a6-9ab3-11ea-9d2d-01aa75ed71a1.0001.02/DOC_1&format=PDF

Everett, M. and Everett, E. 2019. Green Woodpeckers feeding on apples. *British Birds* 112: 48–9.

Fauré, C. (2013) Étude et comparaison des chants du Pic vert *Picus viridis viridis* dans le Sud-Ouest de la France et du Pic de Sharpe *Picus viridis sharpei* dans le Nord de l'Espagne. *Alauda* 81: 209–25.

Fauré, C. (2018) Geographical variation of the song and flight call of the Eurasian Green Woodpecker *Picus viridis* in Europe. *Alauda* 86: 215–33.

Fauré, C. (2021) Characterization of the song of the Iberian Green Woodpecker *Picus sharpei*. *Alauda* 89: 47–60.

Feldner J., Rass, P., Petutschnig, W., Wagner, S., Malle, G., Buschenreiter, R. K., Wiedner, P. and Probst, R. (2006) *Avifauna Kärntens – Die Brutvögel*. Naturwiss. Verein f. Kärnten, Klagenfurt.

Florentin, J., Dutoit, T. and Verlinden, O. (2016) Identification of European woodpecker species in audio recordings from their drumming rolls. *Ecological Informatics* 35: 61–70. https://doi.org/10.1016/j.ecoinf.2016.08.006

Florentin, J., Gérard, M., Turner, K., Rasmont, P. and Verlinden, O. (2017) Towards a full map of drumming signals in European woodpeckers. In XXVI International Bioacoustics Congress, abstract volume: 58.

Florentin, J., Dutoit, T. and Verlinden, O. (2020) Detection and identification of European woodpeckers with deep convolutional neural networks. *Ecological Informatics* 55: 101023.

Forejt, M. and Syrbe, R-U. (2019) The current status of orchard meadows in Central Europe: multi-source area estimation in Saxony (Germany) and the Czech Republic. *Moravian Geographical Reports* 27: 217–28. https://doi.org/10.2478/mgr-2019-0017

Forrester, R. and Andrews, I. (2007) *The Birds of Scotland*. Vol. 2: 960–2. Scottish Ornithologist's Club, Aberlady.

Fox, M. (1997) Green Woodpecker visiting hot-air balloonists. *British Birds* 90: 148.

Fransson, T., Jansson L., Kolehmainen T., Kroon, C. and Wenninger, T. (2017) EURING list of longevity records for European birds. https://euring.org/data-and-codes/longevity-list?page=3

Friedmann, V. S. (1993) Some observations on formation of a mixed pair of Grey-headed Woodpecker (*Picus canus* Gm.) and Green Woodpecker (*P. viridis* L.). *Sbornik trudov Zoologicheskogo Muzeia* 30: 183–96.

Friedmann, V. S. (2011) A riddle of the Green Woodpecker: how appear hybrids with the Grey Woodpecker? *Ethology, Berkut* 20 (1–2): 127–38. (In Russian with English summary.)

Fröhlich, A. and Ciach, M. (2020) Dead tree branches in urban forests and private gardens are key habitat components for woodpeckers in a city matrix. *Landscape and Urban Planning* 202: 103869. https://doi.org/10.1016/j.landurbplan.2020.103869

Fuchs, J., Pons, J. M., Ericson, P. G. P., Bonillo, C., Couloux, A. and Pasquet, E. (2008) Molecular support for a rapid cladogenesis of the woodpecker clade Malarpicini, with further insights into the genus *Picus* (Piciformes: Picinae). *Molecular Phylogenetics and Evolution* 48: 34–46. https://doi.org/10.1016/j.ympev.2008.03.036

Fuller, R. J. (1982) *Bird Habitats in Britain*. T. and A. Poyser, Staffordshire.

Gedeon, K., Grüneberg, C., Mitschke, A., Sudfeldt, C., Eikhorst, W., Fischer, S., Flade, M., Frick, S., Geiersberger, I., Koop, B., Kramer, M., Krüger, T., Roth, N., Ryslavy, T., Stübing, S., Sudmann, S. R., Steffens, R., Vökler, F. and Witt, K. (2014) *Atlas Deutscher Brutvogelarten* (*Atlas of German Breeding Birds*). Stiftung Vogelmonitoring Deutschland und Dachverband Deutscher Avifaunisten, Münster. 368–9. (In German with English summary.)

Gill, F., Donsker, D. and Rasmussen, P. (eds). (2022) *IOC World Bird List* (v12.1). https://doi.org/10.14344/IOC.ML.12.1

Ginn, H. B. and Melville, D. S. (1983) *Moult in Birds*. British Trust for Ornithology Guide 19, Tring.

Glue, D. E. (1993) Green Woodpecker *Picus viridis*. In D. W. Gibbons, J. B. Reid. and R. A. Chapman (eds) *The New Atlas of Breeding Birds in Britain and Ireland: 1988–1991*, pp. 264–5. Poyser, London.

Glue, D. E. and Boswell, T. (1994) Comparative nesting ecology of the three British breeding woodpeckers. *British Birds* 87: 253–69.

Glue, D. E. and Südbeck, P. (1997) Green Woodpecker *Picus viridis*. In E. J. M. Hagemeijer and M. J. Blair (eds) *The EBCC Atlas of European Breeding Birds: Their Distribution and Abundance*. Poyser, London.

Glutz von Blotzheim, U. N. and Bauer, K. M. (eds) (1994) *Handbuch der Vögel Mitteleuropas. Band 9. Columbiformes–Piciformes*. AULA-Verlag Gmbh, Wiesbaden.

Goodge, W. R. (1972) Anatomical evidence for phylogenetic relationships among woodpeckers. *The Auk* 89: 65–85. https://doi.org/10.2307/4084060

Gorman, G. (1995) Identifying the presence of woodpecker (*Picidae*) species on the basis of their holes and signs. *Aquila* 102: 61–7.

Gorman, G. (2004) *Woodpeckers of Europe: A study of the European Picidae*. Bruce Coleman, Chalfont St Peter.

Gorman, G. (2010) *The Black Woodpecker: A monograph on* Dryocopus martius. Lynx Edicions, Barcelona.

Gorman, G. (2011) Green Woodpecker excavating a cavity in autumn. *British Birds* 104: 276.

Gorman, G. (2014) *Woodpeckers of the World: The Complete Guide*. Helm, London.

Gorman, G. (2015) Foraging signs and cavities of some European woodpeckers (Picidae): Identifying the clues that lead to establishing the presence of species. In H. Winkler and F. Gusenleitner (eds) *Developments in Woodpecker Biology*. Biologiezentrum des Oberösterreichischen Landesmuseums, Linz.

Gorman, G. (2017) *Woodpecker*. Reaktion, London.

Gorman, G. (2019) Characteristics of Grey-headed Woodpecker (*Picus canus*) cavities in Hungary. *Aquila* 126: 33–9.

Gorman, G. (2020a) Observations of Grey-headed Woodpecker on cliffs and quarry walls: a seasonal shift in foraging habitats. *British Birds* 113: 567–9.

Gorman, G. (2020b) Reverse mounting by three European *Dendrocopos* woodpeckers. *British Birds* 113: 180–2.

Gorman, G. (2020c) Attributes of Eurasian Green Woodpecker (*Picus viridis*) nest cavities in Hungary. *Ornis Hungarica* 28: 204–11. https://doi.org/10.2478/orhu-2020-0025

Gorman, G. (2022) *The Wryneck: Biology, Behaviour, Conservation and Symbolism of Jynx torquilla*. Pelagic Publishing, Exeter.

Gorman, G. and Alder, D. (2022) Substrate influences foraging selection by Eurasian Green Woodpeckers *Picus viridis* in autumn and winter: observations in Hungary over a 20-year period. *Ornis Hungarica* 30: 170–78. https://doi.org/10.2478/orhu-2022-0013

Gorman, G. et al. (2021) Zöld küllő – *Picus viridis* – European Green Woodpecker. In T. Szép et al. (eds) *Magyarország madaratlasza – Bird Atlas of Hungary*, pp. 433–5. Agrarminiszterium, Magyar Madartani es Termeszetvedelmi Egyesulet, Budapest. (In Hungarian with English summary.)

Grangé, J.-L. and Fourcade, J.-M. (2019) Caractéristiques des arbres de nid de la guilde des Picidés des Pyrénées Occidentales et des Landes (Features of nest-trees of the woodpecker guild in southwestern France, western Pyrenees and Landes). *Alauda* 87: 267–82.

Grangé, J.-L., Senosiain, A., Marsaguet, P. and Navarre, P. (2020) Creusement de cavités en automne-hiver par les Picidés européens (Excavation of cavities by European woodpeckers in autumn-winter). *Ornithos* 27: 345–54. (In French with English summary.)

Gregory, R. D. et al. (2007) Population trends of widespread woodland birds in Europe. *Ibis* 49: 78–97. https://doi.org/10.1111/j.1474-919X.2007.00698.x

Grim, T., Kovařík, P., Harmáčková, L., Tošenovský, E., Hladká, T., Spáčil, P., Krištín, A., Poprach, K. & Sviečka, J. (2022) First documented urban breedings of the Eurasian Scops Owl (*Otus scops*) in Czechia. *Sylvia* 58: 17–35. (In Czech with English summary.)

Handrinos, G. and Akriotis, T. (1997) *The Birds of Greece*. Helm, London.

Hägvar, S., Hägvar, G. and Monness, E. (1990) Nest site selection in Norwegian woodpeckers. *Holarctic Ecology* 13: 156–65. https://doi.org/10.1111/j.1600-0587.1990.tb00602.x

Hann, C. (1951) Courtship Feeding of Green Woodpecker in August. *British Birds* 44: 134.

Hansen, W. and Synnatzschke, J. (2015) *Die Steuerfedern der Vögel Mitteleuropas* (*The Tail Feathers of the Birds of Central Europe*). World Feather Atlas Foundation.

Haraszthy, L. (2019) *Magyarország fészkelő madarainak költésbiológiája* (Breeding biology of birds nesting in Hungary). Vol. 1. Non-Passeriformes, 839–41. Pro Vértes, Csákvár.

Harris, S. J., Massimino, D., Balmer, D. E., Eaton, M. A., Noble, D. G., Pearce-Higgins, J. W., Woodcock, P. and Gillings, S. (2020) *The Breeding Bird Survey 2019*. Research Report 726. British Trust for Ornithology, Thetford.

Henderson, A. J. K. and Henderson, A. C. B. (2002) High and low Green Woodpecker nest holes. *British Birds* 95: 88.

Hoffman, H. J. (1951) Green and Great Spotted Woodpeckers occupying the same hole. *British Birds* 44: 282–3.

Hosking, D. (2011) Green Woodpecker regularly visiting thatched roof. *British Birds* 104: 220.

Iankov, P. (ed.) (2007) *Atlas of Breeding Birds in Bulgaria*. BSPB, Conservation Series, Book 10, 360–1. Sofia.

IOC (2017) World Bird Names List, version 7.2. https://doi.org/10.14344/IOC.ML.7.2

Issa, N. and Muller, Y. (eds) (2015) *Atlas des Oiseaux de France métropolitaine. Nidification et Présence hivernale*. Vol. 2, pp. 804–7. Delachaux et Niestlé, Paris.

Ivanchev, V. P. (1993) (A case of hybridization of woodpeckers of the genus *Picus*.) *Sb. Trud. Zool. Muz.* 30: 197–200. (In Russian.)

Janisch, M. (1954) Birds nesting in company. *Aquila* 55–8: 307.

Jenni, L. and Winkler, R. (2020) *The Biology of Moult in Birds*. Helm, London.

Judson, O. P. and Bennett, A. T. D. (1992) 'Anting' as food preparation: formic acid is worse on an empty stomach. *Behavioral Ecology and Sociobiology* 31: 437–9. https://doi.org/10.1007/BF00170611

Jusys, V., Karalius, S. and Raudonikis, L. (2012) *Lietuvos paukščių pažinimo vadovas* (*The Birds of Lithuania*). Lututė, Kaunas.

Kaboli, M., Aliabadian, M., Tohidifar, M., Hashemi, A., Musavi, S. B. and Roselaar, C. S. (2016) *Atlas of Birds of Iran*. Kharazmi University Press, Karaj, Iran.

Kajtoch, Ł. (2017) The importance of traditional orchards for breeding birds: the preliminary study on central European example. *Acta Oecologica* 78: 53–60. https://doi.org/10.1016/j.actao.2016.12.010

Kalyakin, M. V. and Voltzit, O. V. (eds) (2020) (*Atlas of Breeding Birds of the European Part of Russia*). Fiton XXI, Moscow. (In Russian.)

Kear, J. (2003) Cavity-nesting ducks: why woodpeckers matter. *British Birds* 96: 217–33.

Keicher, K. (2007) Vergleichende Untersuchungen zum Nächtigungsverhalten von Grauspecht (*Picus canus*) und Grünspecht (*Picus viridis*) in Ostwürttemberg (Ostalbkreis) (The behaviour of Grey and Green Woodpeckers (*Picus canus, P. viridis*) in their winter sleeping territories). *Ornithologische Gesellschaft Baden-Württemberg* 23: 3–27. (In German with English abstract.)

Ķerus, V., Dekants, A., Auniņš, A. and Mārdega, L. (2021) *Latvijas Ligzdojošo Putnu Atlanti 1980–2017* (Latvian Breeding Bird Atlas 1980–2017). pp. 266–7. Latvian Ornithological Society, Rīga.

Kessler, J. (2014) Fossil and subfossil bird remains and faunas from the Carpathian Basin. *Ornis Hungarica* 22: 65–125. https://doi.org/10.2478/orhu-2014-0019

Kessler, J. (2016) Picidae in the European fossil, subfossil and recent bird faunas and their osteological characteristics. *Ornis Hungarica* 24: 96–114. https://doi.org/10.1515/orhu-2016-0006

Khaleghizadeh A., Roselaar K., Scott D. A., Tohidifar M., Mlíkovský J., Blair M. and Kvartalnov P. (2017) *Birds of Iran: Annotated Checklist of the Species and Subspecies,* pp. 185–7. Iranian Research Institute of Plant Protection.

King, B., King, M. and Speight, M. C. D. (1973) Winter food of Green Woodpecker and association with Starlings. *British Birds* 66: 33–4.

Kirby, V. C. (1980) An adaptive modification in the ribs of woodpeckers and piculets (Picidae). *The Auk* 97: 521–32.

Klosters, J., Wouters, P. and de Veer, W. (2014) Diet of the European Green Woodpecker *Picus viridis* in the Southern Netherlands. *Limosa* 87: 74–81.

Knaus, P., Sattler, T., Schmid, H., Strebel, N. and Volet, B. (2020) The State of Birds in Switzerland: Report 2020. Swiss Ornithological Institute, Sempach.

Kopij, G. (2017) Breeding densities of woodpeckers (Picinae) in the inner and outer zones of a Central European city. *Sylvia* 53: 41–57.

Korodi Gal, I. (1975) Contributii la cunoasterea biologiei reproducerii si hranei puilor la ghionoaia verde (*Picus viridis*). *Muzeul Brukenthal. Studii si Comunicări. Stiintele Naturii* 19: 329–36.

Kovařík, P., Hladká, T., Harmáčková, L. & Grim, T. (2022) Range expansion of the Eurasian Scops Owl in Czechia. *Sylvia.* 58: 3–16. (In Czech with English summary.)

Kramer, D. (2009) Green Woodpecker drumming on metal plate surrounding nestbox entrance. *British Birds* 102: 142.

Labitte (1953) Notes sur la biologie du Pic-vert *Picus viridis. Alauda* 21: 165–78.

La Mantia, T., Buscemi, I., Mingozzi, T. and Massa, B. (2015) Data analysis on extinct and living woodpeckers (Aves Picidae) in Sicily and Calabria (Southern Italy). *Naturalista sicil.,* S. IV, XXXIX (1): 29–49.

Landler, L., Jusino, M. A., Skelton, J. and Walters, J. R. (2014) Global trends in woodpecker cavity entrance orientation: latitudinal and continental effects suggest regional climate influence. *Acta Ornithologica* 49: 257–66. https://doi.org/10.3161/173484714X687145

Ławicki, L., Cofta, T., Beuch, S., Dmoch, A., Sikora, A., Aftyka, S., Czechowski, P., Bocheński, M., Sieczak, K. and Mazgaj, S. (2015) Identification and occurrence of hybrids Grey-headed × European Green Woodpecker in Poland. *Dutch Birding* 37: 215–28.

Leiber, A. (1907) Vergleichende Anatomie der Spechtzunge. *Zoologica* 6: 51.

Lockwood, W. B. (1984) *The Oxford Book of British Bird Names.* Oxford University Press, Oxford.

Lõhmus, A., Nellis, R., Pullerits, M. and Leivits, M. (2016) The potential for long-term sustainability in seminatural forestry: a broad perspective based on woodpecker populations. *Environmental Management* 57(3): 558–71. https://doi.org/10.1007/s00267-015-0638-2

Löhrl, H. (1977) Zur Nahrungssuche von Grau-und Grünspecht (*Picus canus, P. viridis*) im Winterhalbjahr (Winter-foraging behaviour of Grey-headed and Green Woodpecker). *Die Vogelwelt* 98: 15–22. (In German with English summary.)

MacGillivray, W. (1837–52) *A History of British Birds, Indigenous and Migratory.* Scott, Webster and Geary, London.

Mann, P. (2016) Alters- und Geschlechts bestimmung europäischer Spechtarten anhand des AspekteKonzepts (Determination of the age and sex of European woodpecker species using the aspect concept). *Lanius* 36: 44–58. (In German with English summary.)

Martin, G. R. (2021) *Bird Senses. How and What Birds See, Hear, Smell, Taste, and Feel.* Pelagic Publishing, Exeter.

Massimino, D., Woodward, I. D., Hammond, M. J., Harris, S. J., Leech, D. I., Noble, D. G., Walker, R. H., Barimore, C., Dadam, D., Eglington, S. M., Marchant, J. H., Sullivan, M. J. P., Baillie, S. R. and Robinson, R. A. (2019) BirdTrends 2019: trends in numbers, breeding success and survival for UK breeding birds. British Trust for Ornithology Research Report 722. British Trust for Ornithology, Thetford. www.bto.org/birdtrends

Matysek, M., Wyka, J., Binkiewicz, B., Szewczyk, G., Bobak, J., Wierzbanowski, S. and Cichocki, W. (2020) Liczebność i rozmieszczenie dzięciołów Picidae na terenie Tatrzańskiego Parku Narodowego (Abundance and distribution of woodpeckers Picidae in the Polish Tatra National Park). *Ornis Polonica* 61: 32–46. (In Polish with English abstract.)

Miettinen, J. (2002) Age determination in woodpeckers. In *International Woodpecker Symposium.* P. Pechacek and W. d'Oleire-Oltmanns (eds): 127–31. Forschungsbericht 48, Nationalparkverwaltung Berchtesgaden.

Mihelič, T., Kmecl, P., Denac, K., Koce, U., Vrezec, A. and Denac, D. (eds) (2019) *Atlas ptic Slovenije. Popis gnezdilk 2002–2017* (*Slovenian Breeding Bird Atlas*). DOPPS, Ljubljana. (In Slovenian with English summary.)

Mikusiński, G. (1997) Woodpeckers in time and space: the role of natural and anthropogenic factors. Dissertation. Swedish University of Agricultural Sciences. *Acta Universitatis Agriculturae Sueciae, Silvestria* 40. Uppsala.

Mikusiński, G. and Angelstam, P. (1997) European woodpeckers and anthropogenic habitat change: a review. *Vogelwelt* 118: 277–83.

Mischenko, A. L. (ed.) (2017) (Estimation of numbers and trends for birds of the European Russia (European red list of birds). Russian Society for Bird Conservation and Study, Moscow. (In Russian with English summary.)

Mlíkovský, J. (2006) Egg size in birds of southern Bohemia: an analysis of Rudolf Prazny's collection. *Sylvia* 42: 112–16.

Morimando, F. and Pezzo, F. (1997) Food habits of the Lanner Falcon (*Falco biarmicus feldeggii*) in Central Italy. *Journal of Raptor Research* 31: 40–3.

Morozov, N. S. (2015) Why do birds practice anting? *Biology Bulletin Reviews* 5: 353–65. https://doi.org/10.1134/S2079086415040076

Muschketat, L. F. and Raqué, K. (1993) Nahrungsökologische Untersuchungen an Grünspechten *Picus viridis*) als Grundlage zur Habitatpflege. *Beihefte zu den Veröffentlichungen für Naturschutz und Landschaftspflege in Baden-Württemberg* 67: 71–81.

National Trust (2022) National Trust vows to 'bring back the blossom' as new research reveals massive drop in orchards since 1900s. Press Release: https://www.nationaltrust.org.uk/press-release/national-trust-vows-to-bring-back-the-blossom-as-new-research-reveals-massive-drop-in-orchards-since-1900s

Newson, S. E., Leech, D. I., Hewson, C. M., Crick, H. Q. P. and Grice, P. V. (2010) Potential impact of grey squirrels *Sciurus carolinensis* on woodland bird populations in England. *Journal of Ornithology* 151: 211–18. https://doi.org/10.1007/s10336-009-0445-8

Nilsson, S. G., Olsson, O., Svensson, S. and Wiktander, U. (1992) Population trends and fluctuations in Swedish woodpeckers. *Ornis Svecica* 2: 13–21. https://doi.org/10.34080/os.v2.22398

Obuch, J. (2011) Spatial and temporal diversity of the diet of the tawny owl (*Strix aluco*). *Slovak Raptor Journal* 5: 1–120. https://doi.org/10.2478/v10262-012-0057-8

Olioso, G. and Pons, J.-M. (2011) Variation géographique du plumage des Pics verts du Languedoc-Roussillon (Variation of the plumage of Green Woodpeckers in Languedoc-Roussillon [southern France]). *Ornithos* 18: 73–83. (In French with English summary.)

Opdam, P., Thissen, J., Verschuren, P. et al. (1977) Feeding ecology of a population of Goshawk *Accipiter gentilis*. *Journal of Ornithology* 118: 35–51. https://doi.org/10.1007/BF01647356

Pârâu, L. G. and Wink, M. (2021) Common patterns in the molecular phylogeography of western palearctic birds: a comprehensive review. *Journal of Ornithology* 162: 937–59. https://doi.org/10.1007/s10336-021-01893-x

Patrikeev, M. (2004) *The Birds of Azerbaijan*. 1999–2003 edition. Pensoft, Sofia and Moscow.

Pechacek, P. and Krištin, A. (1993) Nahrung der Spechte im Nationalpark Berchtesgaden. *Die Vogelwelt* 114: 165–77. (In German with English summary.)

Perktas, U., Barrowclough, G. F. and Groth, J. G. (2011) Phylogeography and species limits in the green woodpecker complex (Aves: Picidae): multiple Pleistocene refugia and range expansion across Europe and the Near East. *Biological Journal of the Linnean Society* 104: 710–23. https://doi.org/10.1111/j.1095-8312.2011.01750.x

Perktas, U., Gür, H. and Ada, E. (2015) Historical demography of the Eurasian green woodpecker: integrating phylogeography and ecological niche modelling to test glacial refugia hypothesis. *Folia Zoologica* 64: 284–95. https://doi.org/10.25225/fozo.v64.i3.a9.2015

Perrot, P. (2015) Cohabitation entre Chevevhe d'athena *Athene noctua* et Pic Vert *Picus viridis* (Cohabitation between Little Owl and European Green Woodpecker). *Alauda* 83: 77–8.

Petrovici, M., et al. (2015) *Atlasul pasarilor de interes comumitar din Romania*. Societatea Ornitologică Română (BirdLife Romania) and Milvus Group, Romania.

Poma, E. (1999) Pic vert *Picus viridis*, lombrics et abeilles solitaires (Predation on earthworms and solitary bees by a green woodpecker *Picus viridis*). *Aves* 35: 73.

Pons, J.-M., Olioso, G., Cruaud, C. and Fuchs, J. (2011) Phylogeography of the Eurasian green woodpecker (*Picus viridis*). *Journal of Biogeogreophy* 38: 311–25. https://doi.org/10.1111/j.1365-2699.2010.02401.x

Pons, J., Masson, C., Olioso, G. and Fuchs, J. (2019) Gene flow and genetic admixture across a secondary contact zone between two divergent lineages of the Eurasian Green Woodpecker *Picus viridis*. *Journal of Ornithology* 160: 935–45. https://doi.org/10.1007/s10336-019-01675-6

Prisyazhnyuk, V. E. (2012) (Red List of rare and endangered animals and plants, especially protected in Russia), edition 3, part 1. Vertebrates. (*Bulletin of Red Data Book 5/2012*). Red Data Book Laboratory of All-Russian Research Institute of Nature Protection, Moscow. (In Russian.)

Pritchard, R., Hughes, J., Spence, I. M., Haycock, B. and Brenchley, A. (2021) *The Birds of Wales/Adar Cymru*. Liverpool University Press.

Puzović, S., Radišić, D., Ružić, M., Rajković, D., Radaković, M., Pantović, U., Janković, M., Stojnić, N., Šćiban, M., Tucakov, M., Gergelj, J., Sekulić, G., Agošton, A. and Raković, M. (2015) *Birds of Serbia: Breeding Population Estimates and Trends for the Period 2008–2013*. Bird Protection and Study Society of Serbia and Department of Biology and Ecology, University of Novi Sad.

Quine, C. P. and Humphrey, J. W. (2010) Plantations of exotic tree species in Britain: irrelevant for biodiversity or novel habitat for native species? *Biodiversity and Conservation* 19(5): 1503–12. https://doi.org/10.1007/s10531-009-9771-7

Rassati, G. (2005) Limiti altitudinali del Torcicollo *Jynx torquilla* e del Picchio verde *Picus viridis* in Carnia, Canal del Ferro e Valcanale Alpi Orientali, Friuli-Venezia Giulia (Altitudinal limits of the Wryneck *Jynx torquilla* and Green Woodpecker *Picus viridis* in Carnia, Canal del Ferro and Valcanale eastern Alps, Friuli-Venezia Giulia, [north-eastern Italy]). *Picus* 312: 129–31.

Raqué, K. F. and Ruge, K. (1999) The importance of ants in the food of Green and Grey-headed Woodpecker, *Picus viridis* and *Picus canus* and the influence of agriculture on ants. *Tichodroma* 12: 151–62. Bratislava. (In German with English summary.)

Reichholf, J. (2001) Sich-Sonnen beim Grünspecht *Picus viridis*. *Ornithologische Mitteilungen* 532: 50–2.

Reichholf, J. (2006) Rufaktivitaet des Grünspecht *Picus viridis* im Jahreslauf und ihre Bedeutung fuer die Bestandsentwicklung (The annual cycle of calling activity in the Green Woodpecker *Picus viridis* and its significance to population development). *Ornithologische Mitteilungen* 585: 148–54.

Renwick, A. R., Massimino, D., Newson, S. E., Chamberlain, D. E., Pearce-Higgins, J. W. and Johnston, A. (2012) Modelling changes in species' abundance in response to projected climate change. *Diversity and Distributions* 18: 121–32. https://doi.org/10.1111/j.1472-4642.2011.00827.x

Riemer, S., Schulze, C. H. and Frank, G. (2010) Siedlungsdichte und Habitatwahl des Grünspechts *Picus viridis* im Nationalpark Donauauen (Niederösterreich) (Population density and habitat use of the Green Woodpecker *Picus viridis* in Donau-Auen National Park (Lower Austria). *Vogelwarte* 48: 275–82. (In German with English summary.)

Robinson, R. A. (2005) Birdfacts: profiles of birds occurring in Britain and Ireland. British Trust for Ornithology, Thetford. http://www.bto.org

Rockenbauch, D. (2002) *Der Wanderfalke in Deutschland* (The Peregrine in Germany). Band 2: *Brutbiologie, Ernährung und Wanderungen*. C. Hölzinger, Ludwigsburg.

Rolstad, J. and Rolstad, E. (1995) Seasonal patterns in home range and habitat use of the Grey-headed Woodpecker *Picus canus* as influenced by the availability of food. *Ornis Fennica* 72: 1–13.

Rolstad, J., Løken, B. and Rolstad, E. (2000) Habitat selection as a hierarchical spatial process: the Green Woodpecker at the northern edge of its distribution range. *Oecologia* 124: 116–29. https://doi.org/10.1007/s004420050031

Rozgonyi, S. (1999) Zöld küllő (*Picus viridis*) érdekes éjszakázóhelye (Observation of interesting roosting site of green woodpeckers *Picus viridis*). *Túzok* 4: 27–8.

Ruge, K. (1966) Mischpaar von Grünspecht und Grauspecht. *Journal für Ornithologie* 107: 357.

Ruge, K. (2017) Size and structure of the Green Woodpecker's *Picus viridis* habitat in Baden-Wuerttemberg. *Charadrius* 53: 51–4. (In German with English summary.)

Rustamov, E. A. (2015) An annotated checklist of the birds of Turkmenistan. *Sandgrouse* 37: 28–56.

Salomonsen, F. (1947) En Hybrid mellem Grønspætte (*Picus v. viridis* L.) og Gråspætte (*Picus c. canus* Gm.). *Vår Fågelvärld* 6: 141–4.

Sándor, A. D. and Ionescu, D. T. (2009) Diet of the Eagle Owl (*Bubo bubo*) in Braşov, Romania. *North-Western Journal of Zoology* 5: 170–8.

Schmitz, L. and Dumoulin, R. (2004) Hybridation des Pics vert et cendré (*Picus viridis*, *P. canus*) en Belgique. *Aves* 41: 91–106. (In French with English summary.)

Shimmings, P. and Øien, I. J. (2015) *Bestandsestimater for norske hekkefugler*. NOF (BirdLife Norway) rapport 2: 155–7.

Short, L. L. (1971) The evolution of terrestrial woodpeckers. *American Museum Novitates* 2467.

Sibley, C. G. and Ahlquist, J. E. (1990) *Phylogeny and Classification of Birds: A Study in Molecular Evolution*. Yale University Press, New Haven and London. https://doi.org/10.2307/j.ctt1xp3v3r

Sielmann, H. (1961) *My Year with The Woodpeckers*. Readers Union, London.

Simms, E. (1990) *Woodland Birds*. Bloomsbury, London.

Slabeyová, K., Ridzoň, J. and Kropil, R. (2009) Trendy početnosti bežných druhov vtákov na Slovensku v rokoch 2005–2009 (Trends in common birds' abundance in Slovakia during 2005–2009) *Tichodroma* 21: 1–13.

SLU Artdatabanken: Artfakta (2020). https://artfakta.se/artbestamning

Smith, K. W. (2007) The utilization of dead wood resources by woodpeckers in Britain. *Ibis* 149: 183–92. https://doi.org/10.1111/j.1474-919X.2007.00738.x

Snow, B. and Snow, D. (1988) *Birds and Berries*. Poyser, London.

Snow, D. W. and Manning, A. W. G. (1954) Display and courtship-feeding of Green Woodpecker. *British Birds* 47: 355–6.

Solti, B. (2010) A Mátra Múzeum madártani gyûjteménye III. Németh Márton tojásgyűjtemény (The ornithological collection of the Mátra Museum III Márton Németh's egg collection). *Folia Historico-naturalia Musei Matraensis, Supplementum* 5: 1–275. (In Hungarian.)

SOVON (2018) *Vogelatlas Van Nederland (Bird Atlas of the Netherlands)*. Kosmos Uitgerers, Utrecht and Antwerpen.

Speight, M. C. D. (1974) Anting-like behaviour and food of Wryneck. *British Birds* 67: 388–9.

Spina, F. and Volponi, S. (2008) *Atlante della Migrazione degli Uccelli in Italia. 1. Non-Passeriformi*. Ministero dell'Ambiente e della Tutela del Territorio e del Mare, Istituto Superiore per la Protezione e la Ricerca Ambientale (ISPRA), Tipografia CSR-Roma. (In Italian with English summary.)

Spitznagel, A. (1990) The influence of forest management on woodpecker density and habitat-use in floodplain forests of the Upper Rhine Valley. In A. Carlson and G. Aulen (eds) *Conservation and Management of Woodpecker Populations*. Swedish University of Agricultural Sciences, Department of Wildlife Ecology, Report 17. Uppsala.

Stanbury, A., Eaton, M., Aebischer, N., Balmer, D., Brown, A., Douse, A., Lindley, P., McCulloch, N., Noble, D. and Win, I. (2021) The status of our bird populations: The fifth Birds of Conservation Concern in the United Kingdom, Channel Islands and Isle of Man and second IUCN Red List assessment of extinction risk for Great Britain. *British Birds* 114: 723–47.

Šťastný, K., Bejček, V., Mikuláš, I. and Telenský, T. (2021) *Atlas hnízdního rozšíření ptáků v České republice 2014–2017 (Atlas of Breeding Birds in the Czech Republic).* Aventinum, Praha. 258–9. (In Czech with English summary.)

Stenberg, I. and Hogstad, O. (1992) Habitat use and density of breeding woodpeckers in the 1990s in Møre og Romsdal county, western Norway. *Fauna Norvegica Series C, Cinclus* 15: 49–61.

Stenberg, I. (1994) Grønnspett *Picus viridis.* S. 300 i: Gjershaug, J. O., Thingstad, P. G., Eldøy, S. and Byrkjeland, S. (red.). *Norsk Fugleatlas.* Norsk Ornitologisk Forening, Klæbu.

Stenberg, I. (1996) Nest site selection in six woodpecker species. *Fauna Norvegica Series C, Cinclus* 19: 21–38.

Südbeck, P. (1991) Ein neuer Bastard zwischen Grün- und Grauspecht (*Picus viridis, P. canus*). *Ökologie der Vögel* 13: 89–110.

Südbeck, P. and Brandt, T. (2004) Grün- und Grauspecht sind unterschiedlich – manchmal wissen sie es aber nicht. *Falke* 51: 78–81.

Sultana, J. and Gauci, C. (1982) *A New Guide to the Birds of Malta.* Malta Ornithological Society, Valletta.

Svanberg, I. (2013) *Fåglar i svensk folklig tradition.* Dialogos, Stockholm.

Tate, P. (2007) *Flights of Fancy. Birds in Myth, Legend and Superstition.* Random House, London.

Temperley, G. W. (1951) Status of Green Woodpecker in Northern England. *British Birds* 44: 23–5.

Tomazic, A. (2002) Zelena zolna *Picus viridis* Green Woodpecker. *Acrocephalus* 23: 102.

Török, J. (2009) Zöld Küllő *Picus viridis.* In T. Csörgő, Zs. Karcza, G. Halmos, J. Gyurácz, G. Magyar, T. Szép, A. Schmidt, A. Bankovics and E. Schmidt (eds) *Magyar madárvonulási atlasz* (Hungarian Bird Migration Atlas), pp. 385–6. Kossuth Természettár, Kossuth Kiadó, Budapest. (In Hungarian with English summary.)

Tracy, N. (1946) Some notes on the nesting of the Green Woodpecker. *British Birds* 39: 19–22.

Turner, K. and Gorman, G. (2021) The instrumental signals of the Eurasian Wryneck (*Jynx torquilla*). *Ornis Hungarica* 29: 98–107. https://doi.org/10.2478/orhu-2021-0007

Turner, K., Gorman, G. and Alder, D. (2022) The acoustic communication of the Eurasian Green Woodpecker (*Picus viridis*). *Ornis Hungarica* 30: 10–32. https://doi.org/10.2478/orhu-2022-0017

Turček, F. (1954) The ringing of trees by some European Woodpeckers. *Ornis Fennica* 31: 33–41.

Tutiš, V., Kralj, J., Radović, D., Ćiković, D. and Barišić, S. (2013) *Red Data Book of Birds of Croatia.* State Institute for Nature Protection, Zagreb.

Tutt, H. R. (1956) Nest-sanitation and fledging of the Green Woodpecker. *British Birds* 49: 32–7.

Ürker, O. and Benzeyen, S. T. (2020) The importance of endangered Anatolian (Oriental) sweetgum forests for the bird species. *International Journal of Nature and Life Sciences* 4: 14–25.

Ussher, R. J. and Warren, R. (1900) *The Birds of Ireland.* Gurney and Jackson, London.

Varasteh Moradi, H., Sepehri Roshan, Z. and Chamanefar, S. (2018) Habitat assessment of Green Woodpecker (*Picus viridis*) in Golestan National Park using classification tree method. *Journal of Animal Research* (*Iranian Journal of Biology*) 30: 515–25.

Verboom, W. C. (2019) Bird vocalizations: Bird vocalizations: 'laughing' call of a Eurasian Green woodpecker (*Picus viridis*). Memo no: 201902v2. *Juno Bioacoustics*, Winkel, Netherlands.

Vermeersch, G., Devos, K., Driessens, G., Everaert, J., Feys, S., Herremans, M., Onkelinx, T., Stienen, E. W. M. and T'Jollyn, F. (2020) Broedvogels in Vlaanderen 2013–2018. Recente status en trends van in Vlaanderen broedende vogelsoorten. Mededelingen van het Insti tuut voor Natuur en Bosonderzoek 2020 (1) Brussel. https://doi.org/10.21436/inbor.18794135

Vogrin, M. (2011) Overlooked traditional orchards: their importance for breeding birds. *Studia Universitatis Babes-Bolyai, Biologia* 56: 3–9.

Weißmair, W. and Pühringer, N. (2015) Population density and habitat selection of woodpeckers in mountain forests of the Northern Limestone Alps (Austria). In H. Winkler and F. Gusenleitner (eds) *Developments in Woodpecker Biology*. Biologiezentrum des Oberösterreichischen Landesmuseums, Linz, Austria. 113–33.

Wenger, A. (2021a) Green Woodpecker's Tale (1), January 2021. *Avibus – Birding for Trackers.* https://shop.bewandert.eu/?product=12-learning-nuggets-onsubscription

Wenger, A. (2021b) Green Woodpecker's Tale (2), February 2021. *Avibus – Birding for Trackers*. https://shop.bewandert.eu/?product=12-learning-nuggets-onsubscription

White, G. (1906) *The Natural History of Selborne*. Everyman Edition, J. M. Dent, London.

Wilk, T. (2020) *Picus viridis* Eurasian Green Woodpecker. In V. Keller, S. Herrando, P. Voříšek, M. Franch, M. Kipson, P. Milanesi, D. Martí, M. Anton, A. Klvaňová, M. V. Kalyakin, H.-G. Bauer and R. P. B. Foppen. *European Breeding Bird Atlas 2: Distribution, Abundance and Change*. European Bird Census Council and Lynx Edicions, Barcelona. 494–5.

Winkler, H. (2015) Phylogeny, biogeography and systematics. In H. Winkler and F. Gusenleitner (eds) *Developments in Woodpecker Biology*. Biologiezentrum des Oberösterreichischen Landesmuseums, Linz, Austria.

Winkler, H., Christie, D. A. and Nurney, D. (1995) *Woodpeckers: A Guide to the Woodpeckers, Piculets and Wrynecks of the World*. Pica Press, Robertsbridge.

Winkler, H., Gamauf, A., Nittinger, F. and Haring, E. (2014) Relationships of Old World woodpeckers (Aves: Picidae): new insights and taxonomic implications. *Annalen des Naturhistorischen Museums Wien, Serie B (Botanik und Zoologie)* 116: 69–86.

Winkler, R. (2013) *Mauserumfang und Altersbestimmung von Spechten* (*The extent of moult and age determination of woodpeckers*). Sempach. (Ornithol. Informationsbl. der Schweizerischen Vogelwarte). https://www.ala-schweiz.ch/images/stories/pdf/2014mauserbestimmungshilfespechte.pdf

Wirdheim, A. (ed.) (2020) *Sveriges fåglar 2020*. BirdLife Sverige.

Zarudny, N. and Loudon, H. (1905) *Gecinus viridis innominatus* subsp. nov. *Ornithologische Monatsberichte* 13: 49. (In German.)

Index

References to figures and photographs appear in *italic* type; those in **bold** type refer to tables